T0263246

JEFF'S VIEW
on science and scientists

Cover design: Isabella Seiler and Greg Harris
Cover photos: Kamilla Schatz and Ewald R. Weibel
Cartoons: copyright www.pfuschi-cartoon.ch

JEFF'S VIEW
on science and scientists

GOTTFRIED SCHATZ

AMSTERDAM • BOSTON • HEIDELBERG • LONDON
NEW YORK • OXFORD • PARIS • SAN DIEGO
SAN FRANCISCO • SINGAPORE • SYDNEY • TOKYO

ELSEVIER

Elsevier BV
Radarweg 29, Po Box 211
1000 AE Amesterdam, The Netherlands

First published 2006

British Library Cataloging in Publication Data
A catalogue record for this book is available from the British Library

Library of Congress Control number: 2005933583
ISBN–13: 978-0-444-52133-0
ISBN–10: 0-444-52133-X

For information on all Elsevier publications visit
our website at http://books.elsevier.com

Typeset by SPI Publisher Services, Pondicherry, India

Working together to grow
libraries in developing countries
www.elsevier.com | www.bookaid.org | www.sabre.org

ELSEVIER   BOOK AID   Sabre Foundation
             International

Transferred to digital printing 2006

CONTENTS

CONTENTS

FOREWORD

IRECALL THAT ALMOST 20 YEARS AGO, WHILE
working at Stanford University, a group of us
would travel twice a week across the bay to
Berkeley where Gottfried (Jeff) Schatz was giving
invited lectures. His fame in presenting lively and clear
talks at a high intellectual level was strong enough to
attract listeners from far and wide.

Fifteen years later, when I took over as Managing
Editor of FEBS Letter, it occurred to me that one
contribution that could make the journal more lively
and attractive to the reader would be a column discussing current issues in science. When considering
who could provide our readers with thoughts on the
implications of scientific discoveries and the hurdles
that scientists face during their career, Jeff instantly
sprung to mind.

Jeff is not only a brilliant scientist, but is also
involved in cultural activities (e.g. he is a fine violinist) and this, combined with his experience as head

of the Swiss Science and Technology Council, made him the ideal person to give us his views on matters concerning our profession: *Jeff's View*.

It must have been coincidental thinking, because when I asked Jeff to contribute five or six columns a year to FEBS Letters he immediately agreed and told me that this was exactly the kind of opportunity that he'd been looking for. So Jeff became a member of the FEBS Letters team and has added wit, life, and human experience to our journal.

The positive response of our readers to *Jeff's Views* was so strong that the publication committee of the Federation of European Biochemical Societies decided to publish the collection of essays as a book. We are sure that you will enjoy his views and hope that some of the columns may cause you to stop and think for a little longer than usual; and maybe even give one of your own views a second thought.

We wish to warmly thank Jeff for his contribution!

Felix Wieland
University of Heidelberg

ABOUT THE AUTHOR

GOTTFRIED SCHATZ WAS BORN ON AUGUST 18, 1936 in a little Austrian village near the Hungarian border. He grew up in Graz, but spent one year as a high school student in Rochester, NY. After receiving his PhD in chemistry from the University of Graz, he joined the Biochemistry Department of the University of Vienna where he began his studies on the biogenesis of mitochondria, and participated in the discovery of mitochondrial DNA.

After his postdoctoral work with Efraim Racker at the Public Health Research Institute of the City of New York on the mechanism of oxidative phosphorylation and a brief interlude in Vienna, he returned to the USA to assume a professorship at the Biochemistry Department at Cornell University in Ithaca, NY. Six years later, he moved to the newly created Biozentrum of the University of Basel, Switzerland, where he became one of the leaders in the efforts to elucidate the mechanism of protein transport into mitochondria.

His scientific achievements were honored by many prizes and elections to scientific academies, as well as by two honorary doctorates. He served as Secretary General of the European Molecular Biology Organization (EMBO), Councilor of The Protein Society, Member of the Swiss National Research Council, and, towards the end of his career, as President of the Swiss Science and Technology Council. As a student and young assistant professor, he also worked as a violinist in Austrian opera houses. He and his Danish wife have three children.

LETTER TO A
YOUNG SCIENTIST

YOU ASK ME WHAT IT IS TO BE A SCIENTIST? That's a profound question for someone just entering university. Profound questions rarely have general answers, so I must give you a personal one, even though we have never met. I did basic research in biochemistry all my adult life, closed my laboratory five years ago, and now try to figure out what it was all about. I am still not done with my figuring and am not sure I ever will be. But let me tell you what I believe science can give you, what price you will have to pay, and what science can not give you. It would not surprise me if, many years down the road, your answers should be quite different.

Science, like all great things in life, has many faces and can mean different things to different people.

You only see what you know to look for, and scientific training will give you much better eyes. Not because of the facts you learn – they age quickly and you should always distrust them a little. A famous American biochemist said this to a Harvard graduating class: 'Half of what we taught you is probably wrong, but unfortunately we do not know which half.' To a scientist, cramming facts is what practicing scales is to a pianist: there is no way around it, but it's not enough.

Science gives you better eyes because it removes mental blinkers and gives your brain a much bigger playground. Most people never worry about distances smaller than one millimeter – say, a tiny screw – or larger than a few hundred thousand kilometers – the mileage requirement in a frequent flyer program. That's a range of about eleven orders of magnitude. Your thoughts, on the other hand, will easily move from the behavior of a proton (10^{-15} m) to the size of the universe (approximately 10^{26} m) – about 41 orders of magnitude. It will be the same with time. Most of your friends who are not scientists slice time to perhaps one hundredth of a second – which may decide a ski race – and think back to the old Greeks or, at best, the Paleolithic – that's 2500 to 100 000 years ago. You will slice time into femtoseconds (10^{-15} seconds), which may decide a fast photochemical reaction, and when you think way back, it will be the Big Bang – some 15 000 million years ago, or the beginnings of life on earth – about 3800 million years ago. Again, science will put you ahead by 17–18 orders of magnitude. Science will not make you smarter, or wiser, or a better human being, but it will plug you into the brains of many smart people who were there before you. It feels good to stand on the shoulders of giants. If you want to know that feeling, science is for you.

Let me talk some more about numbers because they are the essence of our craft. As a scientist you will instinctively feel what numbers mean, or do not mean. Understanding numbers will be your Ariadne's thread that shows you the way – in science as well as in everyday life. To you, 27.99 will be 28, not 27 – go tell this to the average shopper! If you hear that employees' motivation has increased by 26.67 per cent you will know that anyone claiming such a precision is a fool – or a fraud. And if your local newspaper carries the headline that cadmium in your city's water has increased by 50 per cent, you will not fly into a panic, but will want to know absolute levels and toxicity limits. It feels good to be friends with numbers.

Understanding numbers also means that you respect their mystic borders and do not take the names of 'zero' and 'infinity' in vain. You know that the real world has no zeroes or infinities and will mistrust anyone calling for *zero risk, zero pollution, zero alcohol,* or *zero sex.* It's the same for *infinite resources, infinite patriotism,* or *infinite sacrifice.* 'Zero' and 'infinity' are the catchwords of fundamentalists. Or of fools, but (to quote Mark Twain) I am just repeating myself.

Science also teaches you to avoid numbers when they would make no sense. One can certainly assess scientific performance, students' satisfaction, success in teaching, and sometimes even originality, but no true scientist would do so by numbers. Don't be afraid to protest whenever your country's science managers clobber you with Citation Frequencies, Impact Factors, and similar nonsense. Giving a number to something that cannot be accurately quantified is bad science. Bad science is science's most dangerous enemy; it is the Fallen Angel that seeks revenge. Scientists are not the only ones who understand numbers; bankers, accountants, traders and politicians can also be very, very clever with them. But if you are looking for someone who knows when *not* to use them, go for a scientist.

When you talk about these matters to friends and acquaintances, they will complain that 'scientists are so arrogant – they think they know it all.' That's wrong twice over. First, those wide horizons science opens up will never let you forget how little you know and understand. Second, the natural sciences never give you absolute certainty, as pure mathematics can do. The scientific truth of today may be wrong tomorrow. We scientists try to inch closer to a truth that's still very far away and hope that our inching is mostly in the right direction. You may have 'proven' a theory by 1000 experiments – tomorrow's experiment may still disprove it. Let this not dishearten you. If you read Karl Popper (which you should), you will learn that disproving an accepted theory is the only way to advance knowledge. Preachers, demagogues, psychics, gurus, faith healers – it's they, not the scientists, who know it all and who are untouched by doubt. I bet you that scientists say 'I don't know' much more often than most other people.

The uncertainty of scientific knowledge is not weakness, but strength. The blind faith of the fundamentalist, like any inflexible structure, crumbles at the next earthquake. The scientific vision of the world has dynamic stability. It is not chained to facts, but is a way of looking at them. Most institutions demand absolute faith, but science makes skepticism a virtue. Scientists see the world as it is, and not as they want it to be.

This was the good part. But there are other parts. In giving you the *grand tour* of the castle, I must now show you the kitchen.

Science has always been a communal effort, but its ability to give us technological innovation has transformed it into 'big business'. That's certainly true of biochemistry and other branches of molecular biology, which offer the promise of blockbuster drugs and other medical revolutions. The biomedical sciences have become expensive, busy, manipulative, political, and harshly competitive. Worse yet, their practitioners are being forced to fiddle with the truth. When they describe their work, they must gloss over uncertainties, or their manuscript won't get published. If they apply for grants, they must make wild claims, or they won't get funded. If they write letters of recommendation, they must tell white lies, or their letters will be counterproductive. And if they shoptalk with colleagues, they must hold back information, or they might get scooped.

Today's science is too much dominated by clever efficiency experts with short-term vision. They will tell you that hypothesis-driven research is a thing of the past and that you should go for 'data mining' – the screening of computer-generated data banks; that good research only comes from large 'networks'; and that it is your social duty to 'valorize knowledge'. If you get your first job at a European university, chances are that you will have to take orders from a senior professor and be kicked out after a few years, no matter how well you did. A company laboratory may treat you better at first, but still kick you out at the next restructuring, regardless of your performance. And if you are allowed to stay on, you will soon spend most of your time at your computer, toiling over mind-numbing question-naires, mission statements, or grant applications. Each collaborator you take into your group will, over the years, ask for at least two dozen letters of recommen-dation from you, every trip to a foreign meeting will eat up at least one week of your time, and every com-mittee you join will be at least twice the burden you expect. Very soon the entrance to paradise – the labo-ratory – will be blocked by guardian angels with flam-ing swords. They will also stand between you and your family, your friends, and any other interests you may have. Your battle will be on too many fronts.

Much of this has to do with forces beyond our control, but we scientists are also contributing to the mess. We want to be smart and forget to be kind. We think too much about competition, and not enough about generosity. We go for power, and forget that power and science don't mix. We are so anxious to become famous that we have no time to learn what science is all about. There are too many congresses, committees, evaluations, prizes, honors, and elections to academies. There is just too much noise.

For many of us, there is also loneliness. Memories of it still haunt me. The loneliness of being separated from my research team by a wall of paper; the loneliness when my friends and colleagues disbelieved one of my discoveries; the loneliness at a far-away scientific meeting after I had given a bad talk; of reading a particularly vituperative rejection letter for a submitted manuscript; of facing tensions with my research group; of evenings with colleagues who only talked about themselves; and, more than anything, the loneliness of trying to hear the static-mangled voices of my wife and children over a very, very long-distance phone line.

Yes, science's kitchen can be crowded, hot, and hectic. But it turns out fantastic menus, whose taste is well worth the price.

Those menus, however, are nutritionally unbalanced and will not sate you. Don't forget to supplement them, because science gives you only one view of yourself and the world. For example, there is also the mystic and the artistic view. Having these different options is the genius of our human species; failing to balance them against one another is our curse. There are parts of you that science will neither explain nor satisfy. If you see everything through the eyes of science, your vision will be monocular and lack depth. Tens of thousands of years from now, our descendents may well conclude that our Scientific Age gave us a distorted view of the physical and spiritual world. I do not consider this possibility very likely, but the Adagio of Mahler's Tenth Symphony, a Rilke poem, or van Gogh's last paintings tell me things about myself that science never told me. Art can be a second vantage point that can grant you binocular vision and let you see in three dimensions. Make science your home, but also venture beyond its doors.

At the university they will only teach you how to do science. To become a scientist, you must learn to look at science from the outside and make it the object of your skepticism. Science is more than just a profession; it is a way to look at yourself and the world around you. In the end, you must figure things out yourself.

2

HOW (NOT) TO
GIVE A SEMINAR

SOMETIMES I WONDER HOW MANY SEMINARS I have sat through. My brain says 'several thousand', but my gut says 'zillions'. Let's see: progress reports, journal clubs, faculty seminars, job seminars, The Harvey S. Benefactor Distinguished Lecturer Award, acceptance speeches for prizes, the list goes on and on. When advising large institutions I often heard forty or more seminars in a few long days. Yes, 'zillions' sounds about right. We scientists spend an inordinate part of our life in seminars. And here is the bottom line: most seminars are bad. *Real* bad.

Yet seminars *are* important. As a postdoctoral fellow, I published my results in the best possible journal and then thought of the next experiment. This habit must have died out in the late Paleolithic. Now the scientific literature is exploding and nobody even tries to keep up with it any more. Today you must go out and sell your stuff. To be at the forefront, you must head for the storefront.

I would not even dream of telling you how to give a seminar. Three children, fifteen PhD students, and 84 postdocs have taught me that raising a finger is just as bad as raising your voice. My postdoctoral mentor left me a little wooden plaque that says 'He who always agrees with you cannot be very bright.' Yes, it's sexist, but that's how they did things in those days. The plaque adorned my office and greatly impressed my students and postdocs. When I told them what to do, they thought of the plaque and did the opposite. That's how they discovered great things. So let me tell you how *not* to give a seminar.

Let's start with the basics. Your seminar should not inform, but impress. And don't call it 'seminar', for God's sake. That word is a clunker. Today it's 'road show'.

As with any show, the title matters. It must be trendy and get the adrenalin flowing. 'Signal transduction in the inflammatory response' is precise, scholarly – and, well, scholarly. 'TNF R1, RIP, TRAF2

and FADD in NF-kappa B activation' is more like it. 'This guy is hot stuff, a real deep thinker' your colleagues will say to each other, and flock to your lecture. A hip title is also good: 'Sex, drugs and yeast mass mating' should catch their attention in Europe and at most major centers in the USA, but do check things out before you speak at the Pontifical Academy in Rome, or in the US Bible Belt.

Don't bother with introductions. General background, biological significance, earlier work by others — that's all for the birds. The present is now, so get right down to business. The opener 'When Jack, Mary and I did Westerns with RIP monoclonals, we noticed a couple of strange bands' will immediately grab their attention. Showing these bands on screen will also let you kill the room lights early on and then keep them off for the rest of your talk. Let your listeners relax, particularly if your seminar is right after lunch. There is nothing wrong with an innocent post-prandial nap.

There are still people out there who project glass-mounted slides — through things called 'projectors'! Ughh! Today you 'beam Powerpoints'. Don't check out the electronics beforehand — do it while you speak. They never work right away, so you can show how great you are with computers. While you take your time fiddling with the buttons, your audience can enjoy the

Microsoft® logo on screen and Bill Gates gets a little free publicity. Even he deserves a break once in a while.

Your hosts have paid through the nose for their high-resolution beamer, so you owe it to them to squeeze the last little pixel out of it. When slides ruled the earth, a diagram's complexity was limited by the skill and the patience of the drafts people. But now we are talking twenty-first century, and the sky is the limit. Fill the screen with all you've got – preferably raw data straight out of your lab notebook. Let the audience feel the pulse of discovery. There used to be a rule that said 'No more than one slide every two minutes.' Baloney! Today's generation was reared on TV and video games and is hooked on images. So keep those pictures coming.

Ages ago, lecturers used wooden sticks to point to things on the screen. They don't sell such contraptions any more, because everyone is into lasers – *Star Wars* stuff. They are cool gadgets, so use them. Keep them on, and keep them moving back and forth until the heads of your audience make you think of a tennis match. Later on they will say that your lecture visibly moved the audience. If the laser battery dies, keep on pointing. This will keep your listeners alert, because they must now search for a dot they cannot see on an image they do not understand.

Don't ever look into the audience. Keep your eyes on the action – the screen. If it happens to be blank, your lecture notes will also do. Once a friend of mine did look into the audience and saw so many people dozing that he was marked for life and never lectured again.

Never talk without lecture notes. Leave that to actors, politicians, and other frivolous folk. You are a scientist, an *intellectual*. So act like one and read your talk in the time-honored meter of scholarship – the monotone. If you cannot do without some spontaneity, follow this simple three-step protocol: (a) don't staple the pages of your notes together; (b) drop them on the way to the podium; (c) use them in the order you picked them up. Your talk will be remembered for its startling connections, sure signs of a creative mind.

Your talk should focus on a single point – *you*. Nobody expects you to be a talking edition of *Annual Review of Biochemistry*. All those great ideas – you had them first. It was you who foisted them on your unbelieving collaborators who then did the easy experiments. If you cannot avoid mentioning ideas of others, explain why they are wrong. Your talk can be elliptical, as long as you occupy both focal points. It wouldn't hurt to throw in a little chauvinism.

Competitors from your own country always have full names. Competitors from elsewhere can be taken care of by collective epithets such as 'a couple of Chinese' or 'a bunch of Dutchies'. If you are British, 'Work by the distinguished Sir X at Oxford and by some Europeans' will please those from Great Albion. If you are American, you can refer to most others as 'people from overseas'. And if you are privileged to work in California, it's simply 'The Coast'. We all know there is no other.

Stay away from simple language. Simple words spell simple minds. Even the international language of science, bad English, loves newspeak. No wonder, the two are close cousins. You never 'read journals'; you 'keep abreast of the literature'. You don't 'do good science'; you are 'at its cutting edge'. Your postdocs are not simply 'good'; they are 'the brightest and the best'. You never 'work hard'; you 'seek aggressively'. And experiments are never 'unfinished, inconclusive or a failure', but 'in press'.

Halfway through the talk, your time is usually up. Now is the moment to think of a scientist's three most important goals: (a) the Nobel Prize (b) unlimited research funds; and (c) unlimited speaking time. To get (a) and (b), you must have brains; to get (c), you must have guts. So don't skip anything – say it faster. Give the audience a rousing coda – they know that the coda is always the fastest part of a piece. No matter how much longer you still want to go on, keep saying 'now, in closing' or 'in these last few diagrams'. That's a great way to keep people from leaving.

When you have finished, do not summarize what you have said. Who wants to hear things twice? Get ready for the discussion, because that's where things might get tricky. Without those beamed diagrams you are left out in the cold. And some listeners may turn cranky, because the lights wake up the old geezers who sit in the front, and switching off the beamer sends the young ones into Acute Visual Deprivation Shock. It's wartime. Take every question as an excuse to continue your talk. Don't answer to the point, and make discussants feel dumb for their inane question. If you are cornered, don't say 'I do not know' or 'OK, you are right', but tell them that your many *papers in press* will answer everything. And if you cannot be right, be wrong at the top of your voice.

Let me end with some honest positive advice. On leaving the lecture room, try to duck the usual hand-shakes and small talk. They can get you into hot water, particularly if you remember faces as badly as I do. If somebody looking vaguely familiar traps you with outstretched arms and a familiar grin, try a generic opener such as 'How was the trip?' That's a safe one, because biologists get around. Stay away from inquiring about the spouse because, as I just said, biologists get around. Relax and enjoy your drink while your intimate stranger shares with you his horror story about the canceled flight. But then it is time to go. Do not stay for the official Dean's reception and the dinner. Mumble something about next day's lecture at a famous place, and head for the airport. Being a hit as a lecturer may get your career off the ground, but a hit- and run-lecturer has arrived. Besides, you can get home early and actually write all those *papers in press.*

3

ME AND MY
GENOME

ON JUNE 26, 2000, PRESIDENT BILL
Clinton and Prime Minister Tony Blair
announced to the world that the complete
chemical structure of the human genome had been
worked out. A genome, they reminded their audience,
is all of a cell's genetic information that is written
down as a sequence of nucleotides in threadlike DNA
molecules. During the next weeks the brouhaha in the
news media was mostly about the promises of new
drugs, prescient diagnostics, and higher stock prices.
Few comments acknowledged that the sequence
was still incomplete and that it was above all a
philosophical landmark.

I had always expected that the sequence of the human genome would tell me many new things about myself. But once I had struggled through the weighty issues of *Nature* and *Science* that described the 'draft' sequence, my first reaction was disappointment. I had not thought that this 3.2 gigabyte message would have so little to tell me. I had expected 100 000 genes, but it only showed a mere 30 000–40 000. Now, a few years later, this estimate has dropped to around 25 000. I find it hard to swallow that I have only ten times more genes than those lowly bacteria in my gut. I had always liked the fact that they had about a thousand times less DNA than I did – that felt about right – but a factor of ten was carrying democracy a bit too far.

Perhaps even a factor of ten could explain the leap from bacteria to humans if the critical factor were not the number of genes, but the number of interactions among them. Boost the number of genes tenfold, and the possible combinations go through the roof. Such reasoning might also take care of the disquieting fact that fruit flies and worms have almost half as many genes as I do. Perhaps so, but the answer cannot be just the number of genes. The answer must have something to do with how many proteins these genes can specify. And if it comes to the number of different proteins, I leave bacteria in the dust. In fact, the dust is so dense that I cannot even guess how far ahead I am.

The simplest living cell whose genome sequence we know is *Mycoplasma genitalium*. This creature has only 580 070 base pairs of DNA and must get by with only about 470 protein-coding genes. The proteins resemble the invitees to a very exclusive party. Only the most essential players are invited: enzymes that replicate and express genetic information; a few proteins controlling the correct folding of other proteins; a lot of pumps in the cell membrane for ions and nutrients; and a survival kit for manufacturing ATP – a colorless, water-soluble organic molecule which can drive many energy-requiring cellular reactions and functions as the major biological energy currency. One or two enzymes involved in the metabolism of amino acids have crashed the party, but apart from that there is none of the usual hoi polloi. Like most exclusive parties, this one is a bore: *Mycoplasma genitalium* is condemned to stay attached to more complicated cells because it must parasitize them for essential nutrients it cannot make itself. Even so, it needs all of its exclusive proteins just to survive. Except for rare mutants, all cells of a population are therefore exactly the same. There is no biochemical room to move. If you happen to be *Mycoplasma genitalium*, you better forget about individuality.

Free-living bacteria such as *Haemophilus influenzae* or *Escherichia coli* can breathe a little easier. According to the latest census, *Haemophilus influenzae* has about 1700 protein-coding genes, and *Escherichia coli* K12 has about 4300. Because these bacteria can modify finished protein chains by clipping off pieces or by attaching other chemical groups, the actual number of a cell's different proteins (called the 'proteome') probably exceeds the number of protein-coding genes. We do not know by how much, but the difference is probably less than twofold. Despite the fact that these free-living bacteria have about ten times as many proteins as *Mycoplasma genitalium*, they still need most of them to survive in the wild. In most respects, therefore, the cells of a population (again discounting rare mutants) are identical. But there are exceptions. These bacteria usually have rotating flagellae coupled to chemosensors by which they can swim towards food, or away from poison. The molecular principle underlying this simple yes/no decision is surprisingly similar to that governing the more complex decisions in our own brain. And these decisions are not always predictable. Look at a swarm of *Escherichia coli* in the light microscope and add a drop of glucose solution

to one edge of the cover slip: some cells will immediately start to swim straight at the food, whereas others will have trouble making up their mind, or keeping
a straight course. The cells have the same genes and
the same environment, yet behave differently. When
they look for food, they show some individuality. Not
much to write home about, but still impressive for cells
with only a few million base pairs worth of DNA.

With its 3200 million base pairs, my own genome
is much larger, and immensely more mysterious. Less
than 1.5 per cent of it represents typical genes, and I
have little idea of what the rest is good for. But I do
know quite a bit about how my cells read their genes
to make proteins. That's the department in which they
really shine. They can, of course, read them from the
beginning to the end just like bacteria do, but they
may also start later, finish earlier, or skip sections in-
between. They can perform similar tricks during the
reactions that transform a gene's information into a
finished protein chain. Yet my cells' ingenuity really
takes off once they have finished a protein chain. They
can cut away pieces from either end, or attach to it
an astonishing assortment of chemical groups that
may affect the protein's function, its location within
the cell, its life expectancy, or its interaction with
other proteins. My cells have at least one thousand

proteins whose major, if not only, job it is to hook a phosphorus-containing group onto another protein. The magic wand of these *protein modifications* gives our proteome polychromatic glitter. To add to this glitter, almost each gene in my body cells exists in two copies that may differ from each other. There may also be thousands or even tens of thousands of very small proteins that we are not even aware of, either because our analytical methods miss them or because our computers are not yet smart enough to pick out the corresponding genes within our DNA.

And there is yet another level of complexity. The food-burning power plants (the *mitochondria*) of my cells carry their own small genome that is completely different from the 'human genome' in my cells' nucleus. A cell may contain up to several hundred mitochondria, each of them with up to several dozen copies of this *mitochondrial genome*. Because these copies may differ slightly from one another and because their protein products may interact in still unknown ways with the proteins encoded by nuclear genes, the possible interactions between the proteins my cells can make increases still further. But there is more. By reshuffling and hyper-mutating some of their genes, the cells of my immune system can theoretically make a nearly infinite number of different immune proteins.

And to cap it all, there is evidence to suggest that my brain cells can alter the amount or the properties of some of their neurological switch proteins in response to training or other external stimuli. How many different proteins can I make? It is anybody's guess. A conservative estimate would be around 100 000. My personal bet would be closer to half a million – and that's not counting my immune proteins.

As impressive as the protein spectrum of my cells is, the real marvel is its regulation. I have immensely complex devices for fine-tuning the activity of my genes. Some of these devices bind to regulatory parts of my genes and can act quickly, whereas others alter the DNA structure around a gene and can shut the gene off for a lifetime. These devices are complexes made up of different proteins, some of which sense what needs to be done. If these devices go out of control, disaster follows. The versatility, subtlety, and mind-boggling complexity of my gene regulation far exceed anything in bacteria. The intricate tapestry of our proteome changes constantly, and we are a long way from understanding how changing one component ripples through the entire system.

Bacteria read their genome. I interpret mine. My genome is not a pedantically annotated musical score that leaves the musician little freedom, but a general bass from which I can evoke many different types of music. My genome is so rich because I can read it in so many different ways.

We humans can make so many diverse proteomes that each of us is unique. This even holds for identical twins: Boris Becker's hypothetical identical twin would probably look like his famous brother, but might well be a mediocre tennis player. The immense information laid down within our genome allows our genome to be flexible and grants each of us individuality.

There is no tyrant as merciless as the small genome. It allows neither biological freedom nor individuality. The more information a genome carries, the greater is the possibility for a fertilized egg to give rise to rather different individuals. To me, the information content of a genome ranks an organism in the hierarchy of life. If organisms have biological dignity, then this dignity must be related to genomic information.

Yet a chimpanzee, or even a mouse, has as much DNA and about as many genes as I do. And it is highly unlikely that I owe my humanity to a few key genes. Counting base pairs or genes may be good enough to sketch the tree of life, but not nearly good enough to trace the ramifications of its topmost branches. Some essential feature of my genome still escapes me. I still do not know why I have large frontal lobes, walk upright, and love to play with words. If I want to tell a chimp his place, I cannot (yet) flaunt my genome.

On the other hand, this genome tells me much about where I come from. The distribution of point mutations in neighboring genes from people round the world suggests that I am the offspring of a very small group of humans that split off from a much older African population between 27 000 and 53 000 years ago. Just think of it — all of us Northern Europeans may come from as few as 50 individuals! Was this biological bottleneck caused by the Ice Age, or by a devastating disease? It sure was a close call.

If one multiplies the complexity of a human cell by 10^{13}, the approximate number of cells in our body, it is easy to see that the molecular system of a human being is so complex that we cannot predict its behaviour with any precision. Perhaps we shall never be able to do this, because a system as complex may defy rigorous prediction. We might have so many parts that at least some of our more complex activities are inherently unpredictable. How wonderful! At long last we would know that we are not merely biochemical machines run by a fixed set of grimly determined genes. The complexity of our body would release us from the prison of determinism. It has been argued that a modern airplane with its several hundred thousand parts is as complex as living cells, yet does not behave unpredictably. This is true, even though many frequent flyers may have their doubts. But the comparison is not fair. We humans are not single cells, and the parts of an airplane are invariable, whereas those of our body fluctuate constantly. Airplanes that automatically expand, shrink, or even jettison their parts depending on the flying conditions might indeed be unpredictable, yet safer than present ones.

When scientists say that something cannot be predicted, they usually mean 'not yet' – at least subconsciously. Perhaps our physicists, chemists and mathematicians must still work out fundamentally new ways to deal with things as complex as a human body. Perhaps they will discover new branches of mathematics and physics that apply only to immensely complex systems. We scientists have learned the hard way that an accepted law may break down with systems that are many orders of magnitude bigger, faster, or more complex than those which the law had originally tried to explain. Perhaps a chaos theory a few centuries down the road will be able to deal with human beings and tell us that they are inherently 'unpredictable' – and why.

I am no longer disappointed by the fact that I have so few genes. At least they are not the ironclad laws I had thought them to be. They are open to argument. It feels good to know that my genome is really just a set of rules.

MY OTHER
GENOMES

THE HUNT FOR THE COMPLETE SEQUENCE of the human genome is now a heroic tale. But heroic tales invite hyperbole. I am tired of hearing that the DNA in my cells' nucleus is the *complete blueprint* of what is, or could be, me. There is more to me than that. The sequence of my nuclear DNA is not 'My Genome'. If a genome is an integrated set of biological information that is passed on from one generation to the next, then I have not one genome, but at least three. Maybe more.

Hardly anybody mentions these other genomes, as if they were slightly illegitimate. In fact, that's what they probably are. They are the fruits of illicit trysts that go back a long way.

To start with, there is the DNA in my cell's energy-generating organelles, the *mitochondria*. That's my mitochondrial genome. I should really say 'genomes', because most of my cells have hundreds or even thousands of them. Each of these genomes consists of a small ring of double-stranded DNA that encodes 13 water-insoluble proteins. These proteins are parts of the intricate machinery that burns the food I eat and stores part of the combustion energy as an 'energy-rich' organic molecule which biochemists have nicknamed ATP. My mitochondrial genome only has 16 569 nucleotides and talks a strange lingo. When it uses the three nucleotides 'TGA' it does not mean 'stop the protein chain' like most other well-behaved genomes do, but 'add the amino acid tryptophan to the protein chain'. For 'stop the protein chain', it says AGA, which for a well-spoken DNA would mean 'add the amino acid arginine'. Yet this small genome is nothing to sneeze at. I need every single one of its 13 protein products to stay alive. My mitochondrial genome does not say much, but what it does say counts. It detests small talk.

My mitochondrial genome is also bizarre in many other ways. For example, it is hard to see why it is there at all. To replicate it and to make its 13 protein products, my cells must set aside at least a hundred proteins, all of them encoded by the DNA in the nucleus and made outside the mitochondria. You do not have to be a Swiss banker to wonder why cells put up with such an investment. I once thought that the proteins encoded by mitochondrial DNA are so water-insoluble that they cannot be transported over large distances and must be made right where they are used — inside the mitochondria. This idea is now less attractive to me, because three of my colleagues have genetically engineered yeast cells that can make some of these 13 water-insoluble proteins outside the mitochondria and transport them back in without any obvious problem.

When everything is said and done, there is no logical reason for the existence of my mitochondrial genome. The reason is historical. About 1500 million years ago, some mysterious ancestor cells engulfed respiring Gram-negative purple bacteria and this *endosymbiosis* turned out to be a success. The immigrant bacteria served their host by burning its food and providing it with plenty of ATP, and the host offered a protective environment, or perhaps a more efficient system for safeguarding and replicating the immigrants' DNA. DNA is an easy prey for oxygen radicals coming from the food-burning system in the bacterial cell membrane, and the bacterial immigrants may have been eager to put a safe distance between that membrane and their DNA. Who wants to store precious family records near a sparking fireplace? This DNA transfer may still go on, or may have stopped once the domesticated immigrants started to talk their own genetic dialect. These bacterial immigrants are now my mitochondria and what's left of their genome is my mitochondrial DNA.

That makes two DNA genomes for me. Yet even the two together do not know all it takes to make one of my body cells. This feat also needs information that is not written down in DNA, but in membranes.

My mitochondria arise by growth and division, just as their free-living bacterial ancestors did 1500 million years ago. Most of the mitochondrial building blocks are made outside the mitochondria under the direction of my nuclear DNA, but many of them can only be put together correctly with the help of a mitochondrion that is already there.

Each mitochondrion is a matrix, or template, that tells new building blocks where to go. Large protein complexes or simple viruses can form spontaneously when their parts are mixed together in roughly the correct proportions. For mitochondria this would not work – they are too complex for that. They are made up of about a thousand different proteins and almost as many different lipids, and mixing all these molecules together would lead to a hopeless jumble. So my mitochondria grow by taking up new molecules and when they have reached a critical size, they divide like bacteria. They cling to what is left of their former independence and, although they have been enslaved for ages, still try to keep up appearances. If one of my cells were to lose its last mitochondrion, most of the mitochondrial proteins and lipids would still be made, but would wander around aimlessly and be degraded.

Mitochondria contain at least a dozen 'usher proteins' that help proteins to find their proper place and start working inside the mitochondria. These usher proteins include receptors on the mitochondrial surface that bind incoming proteins, channels that allow these incoming proteins to move across the two mitochondrial membranes, and 'chaperone' proteins that help incoming proteins to fold into their proper shape. All of these mitochondrial usher proteins are themselves imported from the outside. This leads to a 'chicken and egg' situation, because an usher protein can only enter mitochondria and function there if at least one molecule of the very same usher protein is already in place and ready to work. In other words, many usher proteins import and activate themselves. If any of them is inactivated by mutation, mitochondria no longer work as a template, no longer grow and divide, and the cell dies. Once the last template is gone, the lights go out forever.

My cells appear to have a few other membranes that can neither self-assemble from their component molecules nor arise by modification of other structures. One of these membranes is the *endoplasmic reticulum* – an intricate membrane network that fills much of the inside of my cells and manufactures many of my lipids and proteins. *Hydrogenosomes* and *peroxisomes* may be others. And in plants, it is the *chloroplasts*, of course. Each of these membrane-bounded organelles seems to have evolved from a free-living organism that entered symbiosis with another cell, and then became one of its integral parts.

Hydrogenosomes are particularly intriguing. They are rather exotic fellows, because one finds them only in mitochondria-less organisms that eke out a living in oxygen-deficient biological slums. Hydrogenosomes have an uncanny resemblance to mitochondria: they are about the same size, are bounded by two membranes, and share several typical biochemical pathways with mitochondria. Many of their enzymes have very similar chemical structures as the corresponding mitochondrial enzymes, and the biochemical signals that target proteins into hydrogenosomes show a striking resemblance to those that target proteins into mitochondria. Yet unlike mitochondria, hydrogenosomes lack DNA, and some of the primitive nucleus-containing cells in which they exist diverged from the main line of nucleus-containing cells well before typical mitochondria saw the light of day. Most likely, hydrogenosomes and mitochondria arose from a common bacterial ancestor and then went their separate ways. Hydrogenosomes did not need most of the proteins made by present-day mitochondria and may have managed to jettison their DNA completely before this DNA had a chance to adopt its own genetic dialect.

In contrast to hydrogenosomes, peroxisomes are rather ordinary fellows that are present in most higher cells, including my own. Their hull is only made up of a single membrane and they make — and burn — certain fats that protect my cells from oxidative damage. They, too, may have arisen from a respiring bacterium, even though they lack DNA and are in many ways quite different from mitochondria. I wonder how many such licentious encounters are going on at this very moment, spawning future organelles.

My cells' other membrane structures – the Golgi apparatus, lysosomes, endosomes, exocytotic vesicles, the plasma membrane – are all modified forms of the endoplasmic reticulum. If one mistreats cells with drugs, mutations, or other insults, they may lose these derivative membranes without dying, and will reform them from the endoplasmic reticulum if the air is again clear. With mitochondria, and probably also with hydrogenosomes, the endoplasmic reticulum, and peroxisomes, there is no way that this can happen. They must be passed on from one cell to the next. They are part of my biological heritage – my other genomes, as it were. My two DNA genomes can tell me everything about the molecules in my mitochondria, and quite a bit about how these molecules are assembled. But they know nothing about the fact that my mitochondria are enclosed by two distinct membranes, and cannot tell me how these membranes are put together the right way.

That a particular arrangement of molecules can be heritable biological information has been known for almost half a century. The beating hairs (the *cilia*) on the surface of the single-cell animal *Paramecium aurelia* are arranged in characteristic rows such that all cilia of a given row point in the same direction. During sexual reproduction, two *Paramecium* cells attach themselves to each other and form a bridge through which they exchange their genetic material. In most cases, the two partners then separate precisely so that each gets back its normal surface structure. But even a lowly *Paramecium* can get carried away by a flight of passion and occasionally the separation is faulty so that one partner retains a snippet of the other's surface. This surface abnormality is reproduced for many generations in the offspring when these divide non-sexually. DNA does not seem to be involved, because one can get a similar result by surgically grafting a piece of surface the wrong way and letting the disfigured *Paramecium* cell divide asexually. This surface change behaves like a DNA-linked mutation, but is less stable. The inheritance of this acquired trait injects a little Lamarckian heresy into Darwin's theory of evolution and reminds us that nature scoffs at dogmas. The most plausible explanation is that the surface of *Paramecium* is a template for building new surfaces. Reading the DNA of a given *Paramecium* cell could perhaps tell us that there are different *possible*

41

surface structures, but not which of them is actually present.

How much information is there in my cells' membrane structures? I do not know. But I do know that this information is inherited and that it is just as essential as the information that is written down in my two DNA genomes. I could never redraw the borders of Europe's countries by reading textbooks on law or geography, because these borders were shaped by chance skirmishes, clever treaties, or propitious marriages. They are a record of Europe's long and convoluted history. In the same way, I could never reconstruct my cells' membrane borders by reading my two DNA genomes, because these membrane borders are a record of life's dramatic and convoluted history. It is this history that makes everything about them fall into place.

5

THE TRAGIC MATTER

L IVING CELLS ARE THE MOST COMPLEX matter we know in the entire universe. No other matter embodies so much information. It is so unusual because it is so smart. Yet living matter is not only order and information. Apollo may have crafted its fabric, but Dionysus rules its soul. Living matter has a dark side. It fights a losing battle against its own fragility and winds up scorched. It is a tragic matter.

Hold somebody's hand and feel its warmth. Gram per gram, it converts 10 000 times more energy per second than the sun. You find this hard to believe? Here are the numbers: an average human weighs

70 kilograms and consumes about 12 600 kilo-joules/day; that makes about 2 millijoules/gram.second, or 2 milliwatts/gram. For the sun, it's a miserable 0.2 microjoules/gram.second. Some bacteria, such as the soil bacterium *Azotobacter*, can convert as much as 10 joules/gram.second, outperforming the sun by a factor of 50 million.

I am warm because inside each of my body cells there are dozens, hundreds or even thousands of mito-chondria that burn the food I eat. They release part of the combustion energy as heat, but use the rest of it to fuse inorganic phosphate onto the organic mol-ecule ADP. This chemical fusion produces the 'energy-rich' molecule ATP, which drives many of my cells' energy-requiring processes. Billions of tiny fires are burning in me. Yet I am a cold fish compared to the lowly *Azotobacter* bacterium, which has no mitochondria at all, but whose surrounding cell membrane burns food several thousand times faster than I do. A single gram of this passionate bacterium can make – and con-sume – as much as 7 kilograms of ATP per day.

Before evolution hit upon respiration, life's inven-tive genius was crippled by a power shortage. There is no consensus on what happened during life's early stages, because witnesses are hard to come by. But most biologists believe that life started when our atmosphere had little, if any oxygen gas and that cells produced energy through fermentation. We do not know what form of fermentation, but the fermentations in today's cells suggest that this ancient process must have been quite inefficient. As long as life had to make do with this primitive power source, it could not devise more sophisticated organisms because there was no way to give them enough sustainable power. Also, the fermentation fuels, whatever they were, were bound to run out quickly.

But then came a cell that could capture the energy of sunlight. This feat was, and most likely always will be, the greatest event in the evolution of life on earth. Suddenly life had tapped into a nearly inexhaustible energy source – the nuclear fusions in a nearby star. Life's energy crisis had ended and sunlight-harvesting cells exploded over the entire planet. They used light to make ATP from ADP and inorganic phosphate, and to build organic molecules from inorganic ones. But these cells were big-time polluters: when gorging on light, they split water and released a poisonous by-product – oxygen gas. This gas was bad news, because life had not selected its building blocks for oxygen resistance. These building blocks included lipids with carbon-carbon double bonds, nucleic acids, and proteins – all of them sitting ducks for oxygen gas.

Oxygen could destroy them all; it was just a question of time. A much-needed reprieve came from the immense amount of divalent iron that was dissolved in the oceans and that scavenged oxygen gas as insoluble red iron (III) oxides. Life's refuse made the oceans rust, leading to gigantic 'banded iron formations' of Fe_3O_4 on the sea floors, and 'red beds' of Fe_2O_3 all over our planet's surface. But once this oxygen sink was exhausted, oxygen gas started to saturate the oceans and then escape into the atmosphere until it reached its present level of about 20 per cent per volume.

Rarely, if ever, has the success of a single species been such a catastrophe for all the others. The cloud of toxic oxygen gas must have caused mass extinction, but eventually life learned how to protect itself against oxidation and began its epic battle with oxygen gas. It thought up enzymes that converted oxygen-induced disulfide bridges in proteins back to the original sulfhydryl groups; and it found ways to deal with the particularly vicious 'reactive oxygen species': superoxide radicals, hydroxyl radicals, singlet oxygen, and peroxides that form as by-products when oxygen gas reacts with its preys. Cells started to buffer themselves with antioxidants such as α-tocopherol (vitamin E), ascorbic acid (vitamin C) and carotenoids, and invented a battery of sophisticated enzymes, such as catalases, peroxidases, and superoxide dismutases, which could destroy the nefarious 'reactive oxygen species'. And once these defenses held, life mounted a counterattack and used oxygen gas as electron sink for burning organic matter. To do so, it only had to fiddle with something it already had – its clever machinery for capturing sunlight. Now life had respiration, it could make ATP by lighting a fire.

For life, as for humankind, inventing the fire marked a turning point, because it yielded an energy source of unprecedented intensity. Fermentation was too inefficient to make ends meet for anything but the most primitive forms of life. Photosynthesis was sophisticated and efficient, but could never provide intense bursts of power to a small organism – the number of photons hitting earth's surface is just too low. But with respiration, life had a high-performance engine unlike anything before. This engine could be revved up day and night, rain or shine, as long as there was oxygen gas around. In seconds, respiring cells could release energy that sunlight eaters had patiently collected over hours or even days. With this new power source, life could shift into high gear and devise ever more dynamic and complex organisms.

How well would I do with these three power sources during a 150-meter sprint? If I only fermented my body glycogen to lactic acid, I could reach the finish line fast and perhaps without taking a single breath, but by then the accumulated lactic acid would immobilize my muscles and I would probably fall flat on my face.

Solar power? On a sunny day, my body surface receives at best 500 watts of solar energy. Even if I could convert all of it into ATP (something not even plants can do), I would have only about a third of the power needed for a decent sprint. And with my surface of two square meters, I could still catch a lot more photons than a humming bird or a tiny insect.

It is respiration that keeps me going. There is no better example of its power than the flight muscles of insects. They are the Maseratis of muscles and are packed with mitochondria. Thanks to them, a dragonfly can lift twice its own weight, beating our best helicopters by a factor of seven. But as with any high-performance engine, heat is a problem. To avoid overheating, large organisms respire more slowly and have lower levels of respiratory cytochromes than smaller ones. Also, our cells keep a tight rein over their respiration. When they rest and don't need much energy, they slow down respiration. When this respiratory control fails, hell breaks loose.

We know of two – and only two – unfortunate humans to whom this happened. One was a 27-year-old Swedish woman who, in spite of her normal thyroid status, had the highest metabolic rate ever recorded for a human being. She ate prodigiously, yet remained thin and sweated even when exposed to the cold. Mitochondria isolated from small biopsies of her skeletal muscles burned out of control – they respired full blast no matter whether or not ATP was needed. That was in 1959, long before we had the molecular tools to identify defective genes. The doctors could not help her, and she tragically committed suicide ten years later. The second case was a Lebanese woman who vainly sought help at a US hospital. She liked to sit in one of the hospital's cold rooms, having fans blow chilled air at her – a desperate attempt to get rid of the heat from the fires raging in her. She has disappeared and nobody knows what happened to her.

But even a healthy respiration system is not perfect, because it emits sparks. Some of the electrons that should reduce oxygen gas to harmless water leak out from the respiratory system and reduce oxygen gas to the dangerous 'reactive oxygen species' mentioned earlier. These nasty fellows react with almost anything. They oxidize sulfhydryl groups, peroxidize lipids, cross-link proteins and lipids, modify and displace bases from nucleic acids, and break the strands of DNA. They put any soccer hooligan to shame. They hit the DNA in each of my cells between 10 000 and 100 000 times per day – up to once every second. What saves me is a host of enzymes that repair my damaged DNA as well as possible.

Oxidative pummeling is particularly violent with my mitochondrial DNA, because it sits right in hell's kitchen where respiration goes on and most of the sparks fly. To make things worse, mitochondria are not very good at repairing DNA. No wonder that mitochondrial DNA suffers ten times more oxidative damage, and has a seventeenfold higher evolutionary mutation rate than nuclear DNA. This damage also shows in the short run. As I age, some of my mitochondrial genes for making the respiration system get damaged beyond repair. This sets off a vicious spiral because the defective genes now make defective respiratory chains that spark even more and cause still greater DNA damage. Finally, mitochondria can no longer make enough ATP and tell the cell to commit suicide. Much of this happens in my brain and my two retinas, which respire more intensely than most other parts of my body. And if the cells of these tissues die, it's tragedy down the road.

It seems that the ravages of oxygen also make many other things go wrong with me. Should I ever reach the ripe old age of 90, many of the mitochondrial DNA molecules in my brain will have lost big chunks; up to 1 per cent of their deoxyguanosin bases will be oxidized and no longer do a proper coding job; my mitochondria will have ten times less cytochrome oxidase than I had as a boy; and the membranes in my cells will work less efficiently because they are encumbered by oxidized lipids that are cross-linked to proteins. It seems that I have a really bright future.

If you are healthy (and not an experimental chemist), oxidative damage is probably the most serious damage you suffer from the environment. What can you do to protect yourself? Rigorous experiments on this topic are difficult to do, hard facts rare, and consensus is still a long way off. Still, it is probably a good idea to eat lots of vegetables rich in antioxidants and to stay out of intensive sunlight. You might also do a good thing for yourself (and for some pharmaceutical companies) by adding pills of antioxidant vitamins to your breakfast menu. But the best-documented and cheapest way to fight oxygen damage is to eat less. Less food means less respiration, which in turn cuts down on reactive oxygen species. Here we go again – less food. But this time I am not talking waistline – I am talking oxygen. Animals that frequently go into extended low-metabolic torpor, such as the pocket mouse, tend to live much longer than close relatives that are always active. Now you know why university professors have such a long life expectancy.

But sooner or later, oxygen gets us all. It helps us do great things and stay ahead of entropy, but it exacts a steep price. When cells negotiated their covenant with oxygen, they forgot that they are not flameproof. That's life's molecular tragedy. Perhaps our descendants a billion years down the line will negotiate a better deal. But for now, oxygen always wins. Was this the message the myth of Ikaros tried to tell?

MY SECRET
UNIVERSITY

I HAVE BEEN A UNIVERSITY PROFESSOR FOR most of my adult life and never considered myself an endangered species. Nobody is hunting me, and my genome seems to be more or less OK, thank you very much. But now my habitat is shrinking. If universities continue to disappear at the current rate, my future colleagues will have no place to go.

There is no shortage of institutions that have the shingle 'University' over their door, but many of them seem to have a funny notion of this term, or do not give a rusty ruble about truth in advertising.

The concept of our Western university was born in the eleventh century, mainly in Northern Italy and

England, because there were people who wanted to teach and acquire knowledge without ecclesiastic or royal control. Their concept was hugely successful and the universities founded a little later at Paris, Prague, Uppsala, Vienna and elsewhere have decisively forged the cultural and political face of Europe. When Wilhelm von Humboldt, early in the nineteenth century, championed the unity of teaching and research, he put one of the finishing touches on one of the proudest achievements of our civilization.

Those in power have always looked at universities with a wary eye and tried to control them as best they could. As long as the pressure came from the outside, most universities defended themselves well. But when their inner unity started to crumble, matters started to go wrong. Early in the nineteenth century, the 'humanities' started to move away from the 'natural sciences', precipitating an intellectual calamity of far-reaching consequences. Suddenly there were 'two cultures'. To make things worse, society excommunicated one of them. It awarded the humanities the exclusive patent right to *Culture*® and branded the natural sciences as stepping stones to mindless technology and commercial exploitation.

Today, many of our universities are a collection of professional training schools that interact little, if at all, with one another. For example, a typical curriculum in the natural sciences focuses on professional excellence and leaves little room for opening the students' eyes to the limits, the philosophical consequences, or the ethical implications of scientific inquiry. Matters are even worse in the humanities. These have generally fragmented into many highly specialized fiefdoms that are often unfit to give their students a broad vista of our cultural heritage and the intellectual world. We seem to have forgotten that, for a university, lack of diversity is not simply an adversity, but a perversity.

The gulf between the humanities and the natural sciences has weakened both. The humanities are now marginalized, on the defensive, and underfunded; and the natural sciences have been debased as mere engines for technological progress. They are expected to 'valorize knowledge'.

But universities were not designed to valorize knowledge and are not very good at it. On this count, small start-up companies run rings around them. And when it comes to genomic screens for new drug targets or to ruinously expensive clinical trials, universities cannot hold a candle to the big pharmaceutical companies. Today, the general public expects universities to train professionals for the market place – period. Some politicians try to use them as instruments of social change, or to keep unemployment down. And some city fathers love them as a source of revenue. We no longer have a common vision of what a university should be.

Neither do the universities themselves. They should be restless breeding grounds for new ideas, yet have become one of the most conservative of institutions. They should strive to attract and foster young scholars, because these tend to have the best ideas. Yet few private enterprises treat their young staff as miserably as our universities do. And self-administration has become an inverted world in which professors do the administration and administrators decide policy. Universities should be places of science, but only very few on their payroll do science, or care about it. The Rector of a European university once warned me in a stern letter that scientists should not meddle in university politics. A real gem, that letter! If you are looking for a concise summary of what is wrong with our universities, this letter will do nicely.

Yet I am not pessimistic, because I have had the good fortune to work at some great universities and to see what such places can do for you. And I know at least one university that is just about perfect. When despair is closing in on me, I go there to prop up my courage. I bet the place will do the same for you, so let's visit it together.

It is smack in the midst of a big city, yet you see right away that it is a world by itself. The buildings are plastered with posters on every imaginable subject, and the people milling about will impress you more by their liveliness and smart talk than by their sartorial splendor. A big mural in the entrance hall of the Main Administration building tells you what this university tries to do: to give people the knowledge and the courage to think for themselves and to solve problems rationally and with an open mind. The goal is *autonomous* human beings. That's all. Not a word about 'science' or 'profession'. I guess they want to imply that science is just a method, and professional training a welcome by-product. How could you possibly learn to solve problems without doing scientific research? Von Humboldt again. It makes a lot of sense to me.

It's hard to become a professor there and even harder to get accepted as a student. Professors have three official duties: research, teaching, and interaction with the general public. Few professors are good at all three, but they all try. This can be quite a challenge, since many of them speak with a foreign accent. Students are only admitted after careful screening and intensive personal interviews. The interviewers are particularly interested in applicants that do not fit the common mold, and scoff at age limits, quotas and other arbitrary nonsense. They want their university to be a place for unusual people, and know that one can only spot these by talking to them. It's a hard process, but those that get in are proud of it and do their best to succeed.

If you really want to feel the pulse of the place, look at the big displays that cover the other two walls of the entrance hall. One display shows how the present and former students rank each professor's teaching. Everyone loves that display, because it is a sports column, an academic Michelin guide, and a society page rolled into one. The university president, too, pays a lot of attention to this display, and makes sure that the professors know it. The other display shows how the professors scored the latest exams of their students. *Quid pro quo.* It's tough, but fair and keeps up standards.

You may have trouble telling the students apart from the faculty, because the two of them do similar things. Both do research, organize public discussions, and try to learn from each other. Here, too, it's *quid pro quo.* For example, the biochemistry students show their professors how to run computer programs, the latest gizmos for sequencing DNA, or other new tricks. Students and professors work side by side in running the annual 'University Day' for the townspeople, and the 'Open Door' day of their department.

The professors teach what professors around the world are supposed to teach, but also spend a lot of time encouraging their students to do long-term basic research, to go after problems that might be important for technological innovation only several decades down the road. They keep harking back to the same three points: that universities should be places where people still think about what may happen fifty or a hundred years from now; that the short-term mentality of today's society has made such places precious; and that if universities were to capitulate to short-term thinking, one might as well close them down.

The professors have time for all these things because they never go to faculty meetings. In fact, there are no faculties. There are big departments, and various *ad hoc* structures through which different departments work together in order to give their students a broad training. But most of the strategic and organizational decisions are left to a few powerful deans and the university president. If these people misbehave or turn out to be incompetent, the professors have ways to get them fired. But this does not happen very often, and the professors are glad to let competent academic colleagues run the place. That gives them time to do what they became professors for. They know that those who eagerly await the next faculty meeting are rarely the cream of the crop.

The university president is a renowned scholar with a knack for leadership. When they interviewed her and asked about her administrative experience, her answer was a classic: 'None whatsoever – my strongest point'. She got the job and is good at it. Her persuasive powers are legendary. And when they fail and irate professors or unruly students try to go ballistic on her, she can simply stare them down. She also has a good nose for selecting able administrators and a pronounced allergy to the terms 'center of excellence', 'mission-oriented', 'critical mass', 'networks', and 'synergy'. May she live forever.

The student reps do not think much of endless debates on changing the world. They prefer to evaluate their professors and go after those that do a lousy teaching job. Some of them appear on local talk shows that deal with science issues and occasionally one of them even runs for municipal office. They have also persuaded a private foundation to help them operate their own radio station. This station is quite popular because it is irreverently 'green', yet pro-science.

Students decide for themselves what their training should be and must compose their own curriculum. The curriculum needs approval from a professor and should ensure the necessary training for the chosen goal. But it must also include courses – and exams — in areas not directly related to this goal. For example, a biology student could pick archeology, seventeenth-century Serb poetry, or geology — whatever. Playing in the university orchestra or singing in the university choir also counts, but sports do not. They have to draw the line somewhere.

Friends often want to know whether this university is big or small, and whether it is public or private. I have never bothered to look into these questions, because I do not consider them important.

By now it's time to confess that this university is only in my head. If that's not real enough for you, you are dead wrong. It is real enough to help me find my bearings when advising university presidents or governments. It is my professional North Star. It is not a precise blueprint, but a dream, yet without dreams there are no blueprints. My secret university belongs to the world where I meet my parents, both long dead, enchanted moments of my childhood, and teachers that shaped my life. This world grows on me with each passing year, because it holds an ever-larger part of me. Without this inner world, I could not deal with the outside one.

MIGHTY
MANGANESE

WHEN OUR CHILDREN WERE STILL children, I often told them stories in which kind-hearted monsters or beautiful princesses delivered my educational pitches. We professors are supposed to be educators. When I discovered that my students, too, loved stories, I tried to lecture to them in the form of stories. Although kind-hearted monsters and beautiful princesses would no longer do, thrilling discoveries and whacky scientists easily filled the gap. And chemistry — with or without 'bio' — is just made for storytelling. It is so sensual. Which other science enchants you with colors, crystals, smells, and explosions?

If you are looking for a science that caters to your limbic system, pick chemistry and Primo Levi as a guide. Each of the twenty-one tales in his *Il Sistema Periodico* pays tribute to a chemical element – and to the human spirit. There are gripping stories on argon, zinc, nickel, mercury, iron, and on sixteen other elements – but none on manganese. Indeed, when Levi's classic appeared in 1975, manganese was a metal in search of distinction. It meant longer-lasting batteries, better steel, or crush-resistant aluminum cans. Protecting a six-pack of Budweiser was apparently not the topic to fan Primo Levi's imagination. Since then we have learned how profoundly manganese has shaped evolution of life and the face of our planet. Somebody just has to tell its story, so it might as well be me.

Manganese is a cosmic latecomer, a by-product of the life and death of stars. As late as one thousand million years after the 'Big Bang', the universe was still only vast clouds of hydrogen gas hundreds of thousands of light years across. As these clouds collapsed into galaxies and then into individual stars, the immense heat of compression made hydrogen fuse into helium, turning stars into nuclear furnaces. When a star ran out of hydrogen, it switched to helium as a fuel, fusing it into successively heavier nuclei. The bigger the star, the heavier were the nuclei it could form.

That's how the first manganese atoms were born. But once these primary nuclear fires had burned their way up to iron, they ground to a halt because converting iron into still heavier elements does not yield energy, but consumes it. This Iron Curtain spelled the star's doom. Most small stars died peacefully by ballooning into Red Giants, giving off much of their mass (including the manganese they had made) into the interstellar space, and then finishing their days as slowly cooling White Stars. Big stars (and special pairs of smaller stars) made a much more grandiose exit. Their outer layer collapsed into the exhausted inner core, releasing for a few seconds as much energy as all the hundreds of billions of stars in an entire galaxy. The colossal energy release of such a supernova could forge all the 'energy-expensive' elements up to uranium and beyond. It also created lighter elements, including most of our universe's manganese. The manganese-containing cinders of supernovae were blown deep into the galaxy and coalesced again into the next generation of stars. Finally, some of the manganese ended up on our planet Earth.

With each stellar life cycle, the universe became richer in manganese. But there's still only one atom of manganese in about five million others. In my body, manganese accounts for one atom in 50 million. When life picked its building blocks, it selected against manganese. Yet without manganese, life would not have developed the way it did. Early on, life used ribonucleic acid (RNA) to store information and also to catalyze metabolic reactions. RNA is a close relative of our current genetic material, the well-known deoxyribonucleic acid (DNA). Later, living cells entrusted long-term information storage to DNA, which is more stable than RNA, and metabolic catalysis to proteins, which offer much greater chemical and structural flexibility. Lipids were fine for membranes, and long chains and webs of carbohydrates for tough cell walls.

To make all of this work in the long run, cells needed lots of energy, which they could only get through energy-releasing electron transfer reactions. But life's major building materials were not very good at electron transfer. Although some proteins could shuttle electrons by oxidizing and reducing sulfhydryl groups, these reactions were too limited and too slow to quench life's growing thirst for energy. Cells therefore devised a new generation of electron transfer catalysts by spiking proteins with iron, nickel, cobalt, vanadium, molybdenum and, occasionally, manganese.

These metals quickly and reversibly take up and give off electrons in many different ways. None of them is as clever in this department as manganese. Manganese is the only element that can assume up to eleven different valence states, and the colors of its different compounds cover the entire visible spectrum. Molybdenum, nickel, iron, vanadium or copper are electronic whizzes in their own right, but manganese bests them all. When it comes to trading electrons, manganese is the champ.

Life seems to have taken a while to notice this electronic talent, because it used mostly iron, molybdenum, vanadium, or nickel to build its early generation of high-tech metal-containing enzymes. These could do amazing things. They reduced sulfur, carbon dioxide or nitrate with hydrogen gas, or converted nitrogen gas to ammonia. If manganese was used at all, it was mostly put into electronically dull enzymes that broke down sugars or polypeptide chains.

All this changed when some early sunlight-eating bacteria built a manganese-protein complex that used sunlight to wrest electrons from water. The bacteria did not design this complex from scratch, but tacked it onto the chlorophyll-based solar power plant they already had. It took biochemists a long time to identify this water-splitting complex and to work out its composition and three-dimensional structure. It contains four manganese atoms, some calcium, chloride and perhaps also bicarbonate, and four proteins. The protein that binds the manganese atoms also binds chlorophyll and is a key part of the machine that converts sunlight into biologically useful energy – an ion gradient across a membrane. The other three proteins are in close vicinity to the four manganese atoms, but it is not yet clear what they do. The four manganese atoms successively collect the electrons extracted from water and plug them back, one at a time, into four chlorophyll molecules from which light had knocked off an electron. This step resets the solar power plant for another round of light capture. It is a fantastic machine. If you can think of one that's more impressive, please let me know.

Because the extraction of electrons from water releases oxygen gas, this solar-powered manganese complex changed our planet's face. The face became bluish – and oxidizing. That spelled big trouble, because life had devised its enzymes for a reducing habitat, and now all bets were off. Many of the enzymes with built-in molybdenum, nickel or vanadium could not withstand the oxidative onslaught and cells that depended on them either perished, or had to take refuge in anaerobic biological niches. Some of them coped with the new oxygen-containing atmosphere by surrounding their oxygen-sensitive enzymes with oxygen traps, such as oxygen-binding proteins. These were exceptions, though. With the dawn of the Oxygen Age, many of the sophisticated enzymes containing nickel, molybdenum or vanadium became molecular clunkers.

But the metal that had caused all this turmoil also helped life to cope with it. Threatened by oxygen gas and its toxic reaction products, the 'reactive oxygen species', cells built a manganese enzyme that could destroy a particularly noxious oxygen by-product, the superoxide radical. The manganese-containing *superoxide dismutase* converted superoxide radicals to less toxic hydrogen peroxide, which other enzymes (some of them having, once again, manganese in them) then broke down to water and oxygen gas. Cells had devised a superoxide dismutase already during the Anaerobic Age, but had equipped it with iron instead of manganese. The iron enzyme could apparently deal with the occasional oxidative stress that cropped up even during anaerobic life, but now, as things were getting tough, it was no longer up to snuff. Replacing iron by manganese yielded an industrial-strength enzyme, which became one of life's key defenses against oxygen. Most of today's cells cannot do without it, except those that only grow in the strict absence of oxygen. When some respiring bacteria penetrated into other cells and developed into today's mitochondria, they held on to their manganese superoxide dismutase. This enzyme still protects our mitochondria from oxidative damage and mice that lack it die as early embryos.

Bacteria are still trying to invade our body cells, and when they succeed, we are in trouble. Once inside our cells, these unwelcome guests are usually protected from antibiotics, making intracellular bacterial infections a doctor's nightmare. *Mycobacterium tuberculosis*, which gives us tuberculosis, *Mycobacterium leprae*, which causes leprosy, and *Legionella pneumophila*, the culprit of legionnaire's disease, are all such intracellular invaders. Our cells try to fight back, of course, and none more valiantly than our macrophages, special cells of our immune system. They engulf the bacteria and hold them captive in a kind of cellular stomach, the *phagosome*. The membrane around the phagosome has a battery of enzymes that generate reactive oxygen species and direct them at the entrapped bacteria. Luckily for us, this molecular flame-thrower usually kills the bacteria. Unluckily for us, some of these have learned to hide behind a firewall: they secrete large amounts of a manganese-containing superoxide dismutase into the phagosome and destroy the oxygen radicals that are coming at them. The enzyme is normally inside bacteria and how these pathogenic strains export it is still a little mysterious. Without this firewall, the invaders would have little chance: mutations that inactivate either the bacterial superoxide dismutase or its special export system diminish or abolish the bacterium's survival inside our macrophages.

Macrophages seem to know all this, and try to torpedo the building of this firewall. Although they cannot block the activity or the export of the bacterial superoxide dismutase, they can stop its synthesis by starving the bacteria for manganese. Manganese, like all metals, cannot simply diffuse across biological membranes, but must be ferried across them by specific transport proteins. We know some of these transporters, but many are still unidentified.

One of them was discovered when scientists tried to understand why some Indian subpopulations were more resistant to tuberculosis than others. They traced the resistance to a slight variation in the sequence of amino acids of a protein that is found in macrophages. Scientists, being what they are, christened the protein 'natural resistance-associated macrophage protein 1', or 'Nramp 1'. Others then went on to show that Nramp 1 resides in the macrophage's phagosome membrane and that it transports manganese across membranes. Now the pieces of the puzzle fell into place, revealing a mortal combat between macrophages and bacterial invaders: macrophages use their manganese-transporting protein to pump out manganese from their phagosome so that the entrapped invaders have no manganese for building their molecular firewall. But the bacterial invaders also have manganese transporters (some of them similar to our Nramp) with which they try to suck up manganese from the phagosome. This tug-of-war between manganese transporters may decide whether or not a human being will contract tuberculosis. In the course of human history, this cellular battle for manganese has probably cost more human lives than all the national battles for silver and gold.

My life as a biochemist has never dimmed my enthusiasm for the inorganic world. The intense and virtually indestructible colors of cadmium sulfide or mercury oxide, the knife-like odor of chlorine gas, or the lurid glow of a sulfur flame always make my heart beat faster. They are primeval, zero baselines that let me feel the privilege of my own existence.

The emotional force of manganese struck me in full a few years ago when I was flying across Northern Canada in late spring. Most of the snow was already gone, and the vast sweeps of land far below seemed to be colored by a gigantic brush: blue-green run-offs of copper carbonates and mountain ranges brilliantly red from iron oxides that had formed from iron metal and iron salts when oxygen first appeared in our atmosphere. And then there were black deserts of manganese oxides which, even from my comfortable airline seat ten kilometers up, had an ominous aspect. They stood guard over immense amounts of manganese that lay imprisoned at concentrations at least 10 000 times higher than those in my body. A chunk the size of my little house could easily supply all the manganese for our planet's annual production of biomass. If I ever saw a sleeping giant – there it was. Mighty manganese! After iron oxide, jet engines, and me – what are you up to next?

8

MY TWO BLUES

I HAVE ALWAYS LOVED BLUE, BUT NOW MY favorite color is the blue I see with my left eye. I am also the happy owner of a right eye, but matters turned messy when surgeons replaced my cloudy eye lenses with sparkling pieces of polyacryl-amide. That worked like a charm, but my right eye acted up and needed still other high-tech interventions. I came out of all this in one piece, but with two blues. On a cloudless day, my left eye shows me a blue sky tinged with violet, whereas my right eye shows me one with a touch of gray. I wish I could explain it to you better, but how can I describe color? I might as well try to describe to you the taste of my favorite type of apple strudel.

My color schizophrenia is not without its perks. It always reminds me that color, unlike shape or texture, is not an inherent feature of objects, but my way of sensing how they reflect or filter electromagnetic waves. Another bonus is my belief that only I can fully appreciate our granddaughter's striking blue eyes – although I expect that some young man will soon challenge me on that.

I can see only a tiny sliver of the immense spectrum of electromagnetic waves that bathes the universe. This spectrum spans some 16 orders of magnitude – from the 10 kilometer-long radio waves used by our military all the way down to 10^{-12} m-long γ-rays emitted by disintegrating atoms or exploding stars. Life deals mostly with the wavelengths between 300 and 1000 nanometers – a nanometer (nm) being 10^{-12} m. This spectral range includes the ultraviolet (below 400 nm), which we humans, unlike some insects, cannot see; the spectrum from blue to green to red (400 to about 750 nm), which we and many other organisms perceive as light; and the infrared (above 800 nm), which some animals can see, but we feel only as heat.

Blue may have been the first color life saw. Before cells came up with the ingenious trick of converting sunlight into chemical energy, sunlight was a threat to them because of its harmful ultraviolet rays. To avoid it, early single-cell organisms (the *archaea*) developed a blue-light sensor, which controlled the cells' swimming apparatus so that the cells could swim away from blue light. This sensor has two parts. One is the colorless protein *archaeo-opsin*, whose chain of about 250 amino acids is firmly stitched into cell membrane, spanning it seven times. The other part is a yellowish small molecule which chemists call *all-trans-retinal*, but which is basically a close cousin of vitamin A. It is firmly attached to one of *archaeo-opsin's* membrane-spanning regions. The protein with its retinal attached is called *archaeo-rhodopsin*. When blue light hits archaeo-rhodopsin, it pushes a hydrogen ion (a proton) within the protein from one place to another and changes the protein's shape. In a domino-like effect, this shape change ripples through neighboring proteins, which eventually transmit the light signal to the cell's swimming apparatus.

The light-driven proton movement *within* this blue-light sensing *archaeo-rhodopsin* may have inspired cells to convert the sensor (or an evolutionary ancestor) into an energy-capturing solar cell. By fiddling with the arrangement of amino acids in the protein chain, they shifted the absorption peak towards orange, closer to the sun's maximal energy output on earth's surface. They also made the light push protons *out* of the *archaeo-rhodopsin* and all the way *across* the cell membrane to the cell's outside. Because the cell membrane is an electric insulator, the positively charged protons were trapped outside the cell, capturing light-energy as a gradient of protons across the cell membrane. Cells probably already had a membrane enzyme that could pump protons *out* of the cell by breaking down the cell's major energy-carrying molecule, ATP. By working in reverse, this enzyme could now make ATP by letting protons flow *back into* the cell. Coupling these two sleek machines allowed cells to convert the energy of sunlight into the chemical energy of ATP.

Perhaps I am telling the story backwards. Perhaps the proton-pumping *archaeo-rhodopsin* came first, and the blue light-sensing variety came later. Comparing the structures of the two rhodopsins does not reveal a clear-cut genealogy, but in either scenario, cells faced a tricky problem: they wanted orange light to power their proton pump, but did not want too much noxious blue light. To solve this problem, they modified the blue-light sensor so that it absorbed best in the orange region of the spectrum and could steer them towards orange light. However, once the protein had seen orange light and had helped cells to swim towards it, it turned into a blue-light sensor, which could warn cells to dive for cover when there was too much blue light. Life had invented color vision.

Proton-pumping *archaeo-rhodopsin* is one of the most ingenious devices life ever invented. Why did it not develop this device further and put it into all the modern light-capturing organisms of today? Perhaps the machine was too simple. It could not furnish the reducing power cells needed to produce their organic building blocks from carbon dioxide and water, and its light-capture was not all that efficient. Life is always on the prowl for better things, and when it stumbled upon chlorophyll, it held on to it. Chlorophyll absorbed sunlight even better than retinal and also allowed the evolution of systems that extracted reducing electrons from water. Retinal pioneered biological light capture, but chlorophyll walked away with it.

Modern cells draw their energy mostly from respiration or from chlorophyll-based light-capture, or from both. They no longer have much use for proton-pumping *archaeo-rhodopsin*. Still, the protein persists in today's archaea and in many marine bacteria where it backs up the more modern chlorophyll-based light capture. Good old *archaeo-rhodopsin* may still scoop up as much as one-fifth of the light that feeds our oceans' bacteria.

Light-sensing *archaeo-rhodopsin*, however, was headed for bigger things. In its relentless quest for vision, life tested the protein successfully as light sensor in some higher algae and molds. These cells respond to light, whereas mutants lacking the protein do not – they are blind. As modern cells became more sophisticated, they slightly changed the retinal as well as the protein (or one of its molecular ancestors) so that the resulting product (*rhodopsin*) could interact with the sophisticated signal transmitting system through which modern cells monitor their environment. These changes altered the *archaeo-rhodopsin* almost beyond recognition; only the tell-tale seven transmembrane spans of the modern protein (*opsin*) still reveal its archaeal roots.

Once life had seen color, it was hooked on it. The large genomes of modern cells were exciting new playgrounds for experimenting with ever better methods for color perception and cells exploited these playgrounds to the full. Like the archaea before them, they developed a two-color vision system by attaching a retinal to two slightly different variants of the modern opsin. Already more than 800 million years ago, this modern rhodopsin system allowed animals to distinguish blue (below 500 nm) and yellow (above 500 nm). Later on, insects and higher animals duplicated the gene for the yellow sensor and then changed one of the two copies, so that the changed copy responded to red or green. Further modifications of rhodopsin led to systems that could see still more colors: bees, many fish, reptiles and birds can distinguish four colors, and many butterflies as many as five — from deep red all the way into the ultraviolet. Some animals can even peer into the infrared.

In order to see in dim light, most animals also acquired yet another type of rhodopsin that is extremely light sensitive, but cannot distinguish between different colors. About 400 million years ago, many animals had three to five different color sensors as well as a rhodopsin for dim light. Early mammals that hunted mostly at night did not need to see many colors and regressed to two-color vision. It was only 35 million years ago that ancestors of primates and humans re-invented mammalian three-color vision, perhaps because it helped them distinguish ripe from unripe fruits against a background of confusing foliage. Seeing only two colors would have made this task quite difficult. There must be many other amazing things we still do not know about how animals see color. In fact, the only systems we really understand in detail are our own and those of primates.

My retinas have three types of cone-shaped photoreceptor cells that have broad and overlapping absorption peaks in the blue, green and red regions of the visible spectrum. That makes me a *trichromat*. The combination of the relative signal strengths from these three photosensors lets me see more than two million colors. Humans and primates are the only mammals that can see so many colors. Most other vertebrates with their two-color sensors can distinguish only about 10 000 colors – about an order of magnitude less than the screen colors of our latest mobile phones. My color receptors are very good at distinguishing colors and fine detail, but need lots of light. When it gets dim, they go into sleep mode and leave the field to my rod-shaped photoreceptors that are very light-sensitive, but wake up very slowly and give me only low resolution and no more than 200 shades of gray.

Have you ever wondered why many bars and restaurants are so dimly lit? Customers love it because it hides their facial wrinkles and gray hair. Within my retina, rods and cones form an uneven mosaic, and the cones compare the color signals among each other before sending their joint report to my brain for final analysis. Both tissues work very hard at it: my retina consumes more energy per gram than any other part of my body, and my brain is not far behind. That's why my retinas and my brain are particularly prone to go bad with age. Because they are so energy-hungry, they ferment glucose to lactic acid even when there is plenty of oxygen around. No other part of my body wastes glucose like that. At least that's what I hope, because the only exceptions to this rule are cancer cells.

I have about ten to twenty times more rods than cones. My retinas' center is exceptionally rich in shape-discriminating cones and I rely on it for my most acute vision. One of my colleagues has once called the retina's center the most valuable square milli-meter of the human body. This was before young ladies started to stud their navels with diamonds, but I presume his statement still holds true.

My gene for blue-sensing opsin sits far apart from my other opsin genes on chromosome seven. But my genes for the red-sensing and green-sensing opsin sit next to each other on my single X chromosome, because they arose by duplication of an ancient gene for a yellow sensor. As all opsin genes are very simi-lar, this spells trouble for women with their two X chromosomes because two similar genes that are near each other in the genome often exchange parts with one another during sexual reproduction. When women produce egg cells, they occasionally replace the single gene for red-sensing opsin by two genes for the green-sensing opsin on the same X chromosome, or vice versa. The egg will then have an X chromosome that encodes only two green-sensors, or only two red-sensors. That's no problem for the daughters, because they still get the missing color sensor from their father's X chromosome, which is likely to be normal. But the sons with their single X chromosome have a 50 percent chance of being left with only two green sensors, or only two red sensors. Even though the two greens or the two reds will not be identical, they are usually so similar that these unlucky fellows will have only two color sensors: for blue and red, or for blue and green. We call them color-blind, but they are really *dichromats* who see fewer colors than the rest of us.

'Color-blindness' was clearly described only in 1777 by Joseph Huddart in his classic *An account of persons who could not distinguish colours*. Two-color vision afflicts 8 per cent of Caucasian men, but hardly any women. Keep that in mind when you prepare the colored power points for your next plenary lecture, because about 80 males in an audience of 2000 will see them quite differently than you imagine. Yet being a dichromat also has its upsides. The US army likes to pick 'color-blind' recruits as sharp-shooters or scouts because they are not as easily fooled by multi-colored camouflage. For your plenary lecture this means that those 80 male listeners won't be fooled by your multi-colored power points and aim straight at any weak data.

The hanky-panky among the neighboring genes for red-sensing and green-sensing opsin during egg cell formation can sometimes create an X chromosome that encodes two green-sensors that are so different that they should work like two distinct color sensors. A daughter inheriting this X chromosome should then have four distinct color sensors: the blue-sensor encoded on chromosome seven, two different green-sensors from the mother's messed-up X chromosome, and a normal red-sensor from the father's normal X chromosome. But could she plug that extra green sensor into her neural network and actually see more colors than we normal mortals? Would she be a *functional* tetrachromat?

It seems so. When the gene for the human red-sensing opsin is genetically engineered into mice (which only have two color sensors and lack a red sensor), the human sensor appears in the animal's retina and responds to its appropriate color – at least that's what we can tell from testing the retinas by electrophysiological methods. So it seems that the retina of a tetrachromatic woman would make use of the extra sensor. But would her brain know what to do with this additional information?

About ten years ago, British scientists set out to hunt for 'Ms Tetrachromat' among 14 mothers whose sons were color-blind because they had inherited either two green-sensors or two red-sensors. The women were asked to mix red and green lights with a joystick-controlled device in order to recreate a particular hue of yellow-orange that was outside the working range of the human blue-sensor. As expected, normal trichromats, having only their red- and green-sensors to go on, found many matching combinations. But one woman (the 57-year-old *Ms. M*) was exceptionally fussy and found only a single match that satisfied her. According to genetics, she had an extra green sensor that was most responsive to a color between green and red, and apparently she used it well. Still, the existence of tetrachromatic women is not yet solidly established, because it is so hard to do conclusive experiments. We scientists may not be normal people, yet most of us are still trichromats and have no objective way of telling whether a test subject's choice of color match is 'correct'.

What would life be like for a tetrachromatic woman? At times quite a pain, because most photos, movies or TV screens would show her wrong colors. But she might well be exceptionally good at playing computer games, surfing the Internet, or analyzing colored diagrams because she might feed her brain color-coded information through four, rather than three channels. We dumb males could only watch her in awe. But before we decide to marry her we should remind ourselves that her sons would have a 50 percent chance of being color-blind.

Many people have slightly abnormal color sensors that may affect their color perception. The differences are usually minor, but there is no doubt that many of us see colors in a very personal way, which we cannot share with others. When it comes to seeing color, each of us is very much alone. The known combinations of different opsin variants with retinal can only produce color sensors covering the spectral range from 345 to 610 nm. But rare mutations that change opsin, retinal, or some of the molecules that transmit color information to the brain could well extend this range. Such rare human mutants might have exceptional night vision, or be 'mind readers' because they can perceive minute fluctuations of other people's skin color.

Why do I see two different blues? Even when my eyes were still in mint condition, they had less 'blue' cones than 'red' or 'green' cones — like all human eyes. The ratio of blue cones to red and green cones was particularly low in my retina's center because nature tried to compensate for the chromatic aberration of the natural lenses she presented to me at birth. I suspect that the operations on my right eye destroyed too many of the rare blue cones in the retina's periphery. Surgeons would have to open up my eyes to make sure – but that's the last thing I want them to do. I am perfectly happy with the way things are – including my two blues.

Ever since Isaac Newton published his 1671 classic *New Theory of Colours*, our perception of color has intrigued many of our greatest scientific minds. When light proved to be just one form of electromagnetic waves, it still remained puzzling why we cannot create some colors by mixing the others. The existence of these 'primary colors' seemed to imply that the electromagnetic spectrum obeyed some hierarchy which, once understood, would give us fundamental insights into the nature of light. But the three primary colors are just the shadows which our three color-sensors project on the continuous immensity of the electromagnetic spectrum. In the end, understanding primary colors and color vision in general has told us much less about the nature of light than about the nature of ourselves.

POSTDOCS

'WHAT HAVE YOU DONE TODAY TO make me famous?' asks the distinguished professor, putting an arm jovially around his postdoc's shoulder. Welcome to the postdoc universe.

Most dictionaries will not tell you what 'postdoc' means. It is newspeak for 'postdoctoral fellow' – somebody who has graduated with a PhD degree and then gone elsewhere for a few years to do research. A scientific apprentice, as it were.

'Postdoc' – the word says it all. There is the vagueness; 'postdoc' specifies neither rank nor duty. There is the lack of status; nobody outside science has ever heard of the word, so it is useless for calling cards, civil service pay scales, or promotion schemes. And there is the no-man's land: the suffix 'post' stands for the past and has the melancholic ring we associate with post-modernism, post-genomics, or post-mortem analysis. A postdoc is neither here nor there. The terms 'undergraduate student', 'graduate student' or 'assistant professor' promise a future; 'postdoc' does not promise anything.

Forty years ago, when I was at that stage, most post-docs were in effect 'pre-assistant professors'. I still think that's what most postdocs should be. Whereas freshly minted PhDs can, and should, consider a wide variety of job possibilities, in academia, industry, banking, law, science administration, or politics, someone who has invested several years in postdoc training should aim for a position involving research. But today's shortage of assistant professorships has made this goal elusive. Postdocs may wind up as research associates, untenured assistant professors, or guest scientists – anything that sounds good and means nothing. Phony euphemisms are always sure signs that there is a problem, and that's true here as well. The postdoc system is in crisis.

The key problem is that postdocs are finding it harder to find a long-term academic job. The collapse of the Iron Curtain and the scientific growth of many Asian countries have swelled the number of those look-ing for postdoc training in the USA and Western

Europe. But back home, long-term jobs are often scarce or non-existent. And they are also getting scarcer in the USA and Western Europe, which are clamping down on their science budgets in order to straighten out their finances or pay for expensive wars. At US universities, the availability of junior faculty positions is also diminished by tenured professors who refuse to retire. Also, the academic job markets in Europe and Japan are notorious for their insider trading and the rarity of independent junior positions. All these problems have led to a massive traffic jam at the end of the postdoc tunnel. Those trapped in the tunnel must do a second or even third round of postdoc training. But a string of postdoc stints looks bad on a professional résumé and makes it even harder to find a long-term research position.

Another problem is the excessive dependence of many postdocs on their supervisor. Dependence breeds abuse, and abuse is on the rise. Successful PhD graduates from many rich countries are generally immune to this problem, because they can count on comfortable postdoctoral fellowships from their home country and therefore pick almost any postdoc mentor they choose. It is *they* who recruit the supervisor, not the other way around. That's why the quality of the postdocs is one of the best indicators of the quality of a laboratory. Institutions in rich countries can also offer postdocs well-paying temporary staff positions. But many postdocs, particularly those from Eastern Europe, India, China and other less wealthy countries, must find a supervisor who is willing to pay their stipend from a personal grant or some 'slush fund'. There is nothing inherently wrong with such stipends; on the contrary, it is only thanks to them that many young scientists can get their postdoc training. But these stipends can be reduced or even stopped at the supervisor's discretion, making the recipient almost totally dependent on the supervisor.

Some institutions grant all postdocs, even those paid through research grants or foreign fellowships, the same social benefits they offer their regular employees. But many do not, and that's where the problem lies. I have seen a supervisor fire an Asian postdoc on three months' notice, even though the postdoc had a young family to support and neither the money nor the position to return home. I have seen postdocs appointed on as little as 50 or 25 per- cent of a regular postdoc salary, even though they had to work full-time. And, yes, I have seen postdocs victimized by cultural arrogance that sometimes bordered on discrimination. Science has never been a perfect shield against chauvinism or intolerance, and the current nationalistic and fundamentalist paranoia has not helped matters. In this poisoned atmosphere, an Arab, Turkish, Indian or Serb passport – or even name – can be a real scientific handicap; and visa offices won't consider it a bonus, either. Raising this touchy matter is politically incorrect and a university official once gave me a dressing down for offending her with such nonsense. 'The lady doth protest too much,' Shakespeare whispered into my ear. Right as always, William!

History has never known a classless society and our scientific community is no exception. Those who think that we do things only the democratic way should have their head examined. Our science establishment is at best a meritocratic oligarchy, and quite often a monarchy. I could even name you a few fields that resemble religious sects. You may think that the pecking order at our universities starts with the tenured professors and continues with untenured professors, postdocs, graduate students, and undergraduates. But that's poppycock. If one considers official rights, legal protection and professional representation, the true power structure is tenured professors, untenured professors, undergraduates and graduates, with postdocs at the bottom.

Many postdocs, particularly those paid through research grants, have no official rights. Students' Associations, fraternities, graduate committees, or Deans of Graduate Studies do not feel responsible for them. Neither do faculties. Many postdocs even lack proper insurance. They may be on foreign soil and trying to cope with strange customs, an unfamiliar language, or raising young children. Their spouse may have interrupted a career for their sake and now feel isolated and frustrated. And money is nearly always a problem. For young couples, the postdoc period can be the most insecure and vulnerable period of their shared life.

On the other hand, insecurity and vulnerability are hallmarks of development and evolution – in yeast, fruit flies, or in human beings. The international flow of postdocs is the bird migration that selects the best, fights inbreeding, and keeps the scientific community healthy. Having done well in two different laboratories and two different countries is a rather objective quality seal and usually opens many doors. And postdocs, together with the graduate students, are the major engine that drives scientific innovation; postdocs and graduate students do most of the experiments and make most of the scientific discoveries. Without them, we professors would have to roll up our sleeves – and then may the Lord have mercy on us all.

Even though we may have struggled during our postdoctoral years, most of us do not remember them as Purgatory, but as Paradise Lost. It was then that we

could do research without having to sit in lectures, cram for exams, toil in committees, or haggle with students about grades. It was then that we made some of our most original discoveries and chose our long-term field of research. And while the job prospects of today's post-docs are no match for those of young business graduates or MDs, they are way better than those of musicians, writers, painters or actors. Creative professions have always exacted their toll. Keep your ears open at the next faculty party and you might hear a normally reserved colleague talk excitedly about the postdoctoral years: 'The boss was a slave driver all right, but the lab was great. Most of the time we were broke and our miserable car crapped out at the worst moments. But I could work in peace and, boy, did I work hard!' 'You were also still young and crisp,' the spouse may add dryly. There is nothing like a spouse to add perspective.

This rough-and-tumble postdoc universe mirrors that of science itself and we should not over-regulate it. I shudder at the thought that the bureaucrats in Brussels, Washington or Tokyo may decide to put the international postdoc system through their regulatory wringer. Science does not need more regional quotas, Centers of Excellence, uniform vacations, or 35-hour work weeks. It needs young minds willing to try new things, to put up with hard work, and to take risks. Science, the great adventure, needs adventurers.

But science also needs prudence and we should use it to keep the postdoc system healthy. If you are a graduate student, start thinking about where to do a postdoc at least one to one and a half years before you graduate. Shoring up a fellowship can take a lot of time and most good laboratories have a long waiting period. Of course you should look for scientific excellence, but don't forget the human angle. Some famous laboratories are snake pits, which can suck the joy out of science and warp you for life. Selecting your postdoc mentor is one of the most important professional decisions you are ever going to make, so do visit the laboratory you are interested in and talk to the students and postdocs, preferably one-on-one and in private. Are they happy with the lab and the supervisor? Would they choose the place again? Do they get the support they need? And what has become of the previous postdocs?

If you are a PhD supervisor, remember that it is one of your most important obligations to help your students select the most suitable postdoc mentor and write an informed, personal and intelligent letter of application. Many of you do not take this duty seriously, even though it is an essential part of a good graduate education.

If you are a postdoc supervisor, remember that a postdoc's presence in your lab is an unspoken plea to you: 'I have left home to work with you, because I respected and trusted you and wanted to learn from you. I am now in your hands – please take good care of me.' It's a touching message; let it sink in and work on you.

And now to you, my dear postdoc. You, too, have obligations. Do you enjoy the facilities of your host lab, yet drag your heels when it comes to maintaining them? Are you one of those who always call in sick when it's time to clean up the cold room? Will other laboratory members looking for advice make a circle around you because you never have time for them? Will you refuse to chip in when your supervisor needs help in grading an exam? Will you be remembered as one of those obnoxious types who is the curse of a group leader? Will you make it a habit to ask for letters of recommendation one day before the deadline? Will you secretly plunder the laboratory when you leave for another position? I could go on, but I guess you get the drift. As a postdoc you are an important member of the laboratory family whose members need each other. Do not abuse this power.

During my years in research I have worked with more than 80 postdocs and I am often no longer sure who overlapped with whom. But I still remember my reaction when I first met each of them face to face. One of them – let me call him Mark – just received a prestigious prize and his photo in the prize announcement opened a floodgate of memories.

When Mark had first introduced himself to me, his face had shown intelligence, motivation, and that youthful irreverence I now see so rarely as a retired professor. I also sensed an appealing touch of insecurity. He had graduated from a top university but, in spite of his talent, had not done well in his PhD thesis. His PhD supervisor had been on the outs with him and written him a lukewarm letter of recommendation. It had been easy to see that he was a gifted young man, whose start in science had been bumpy and who was struggling with self-doubt.

We immediately hit it off, but I felt that a heavy burden had been dropped into my lap. At that time, working with postdocs was still a novel experience for me. I had been a bad advisor to a previous postdoc because I had been too immature to handle her rough edges. Now I was afraid of another failure and acutely felt that choosing a postdoc was just as crucial, and difficult, as choosing a postdoc mentor. Would I do a better job with this applicant? Where – and who – would he be ten years down the line? During that first encounter, my insecurity may have outstripped his. Perhaps that's why it felt so good learning of his success and seeing the photo of his mature and self-confident face – the face of a familiar stranger. My feelings were that of relief, gratitude, affection, and pride, all rolled into one.

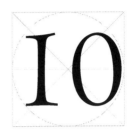

THE ART OF
STEPPING BACK

SCIENTIFIC BREAKTHROUGHS HAVE THEIR own punctuation. Some come with a triumphant exclamation mark because they explain something that had always puzzled us. The discovery of the DNA double helix had *eureka!* written all over it. Other breakthroughs, such as the polymerase chain reaction for analyzing traces of DNA, have an expectant question mark because they give us a new tool that makes our hearts beat faster. And then there are those breakthroughs that come in quiet brackets. It would embarrass me to count those I initially overlooked. When I finally recognize their import, I always feel a little chastened. The claims that the energy of respiration is initially captured as an electrochemical potential across membranes, or that the controlled breakdown of proteins needs an input of energy both made such unobtrusive entrances. The same is true for the discovery that cells can kill themselves.

That life and death go together and even require one another are not new insights. Persephone, who reigned over the dead, was the daughter of Demeter, goddess of life-bringing harvests. But most Western societies see life as active and death as passive. Death is 'suffered' or 'inflicted', be it by age, disease, starvation, or violence. It is outside chaos vanquishing the order of life. Biochemistry and cell biology have reinforced this lopsided view. They showed us that living cells are immensely complex and that cell death by starvation, heat, mechanical injury, or poison wreaks collapse of ion gradients across membranes, swelling of the cell and its mitochondria, general loss of cell constituents, and inflammation of surrounding tissue. Something to be fought at all cost. 'Do not go gentle into that good night!' was the advice of Dylan Thomas.

Life has always faced chaotic death and learned to cope with it. Many organisms produce a grotesque surplus of germ cells or offspring so that a few of them may survive. Evolution has taught these parents that most of their progeny would die a chaotic death, so they try to overwhelm this death by plenitude.

But now we know that there also is another death, that cells can commit suicide in an orderly way. This 'programmed cell death' or 'apoptosis' is neither passive nor chaotic; in fact, it is anything but. Apoptosis shows us death's second face, and it looks remarkably like that of life.

Watching a cell commit suicide is like watching a well-rehearsed ritual. The cell shrinks; its mitochondria disintegrate; the plasma membrane forms blebs and breaks up into small membrane bags – 'vesicles' in biochemist's language – that retain all of the cell's contents and are finally eaten by special scavenger cells, the *phagocytes*. There is neither inflammation nor necrosis. The dying cell neatly seals itself into garbage bags, does not pollute the neighborhood, and avoids public disturbance. It disappears without a noise.

The molecular workings of this hara-kiri process are no less impressive. Apoptosis-regulating proteins form pores in the mitochondrial membrane barriers, allowing small molecules and proteins to escape. Several of these escapees are enzymes that degrade other cell constituents and that also trigger an intricate series of reactions that finally dismantle the entire cell.

Somewhere along the way, the lipids in the vesicles' membrane get jumbled so that some that are normally hidden are suddenly exposed. For roving phagocytes, these exposed lipids are what blood is for roving sharks — a powerful bait. It makes the phagocytes home in on the vesicles, devour them, and release anti-inflammatory signals that keep everything quiet. This basic machinery has been remarkably conserved from simple worms (in which some of its parts were first discovered) to man. All higher forms of life seem to have it. Primitive single-cell organisms with a cell nucleus, such as the common baker's yeast, generally lack the full-fledged machinery, except perhaps for some that are parasites of more complex organisms. Nobody has so far found apoptosis in a bacterium.

There are many variations, footnotes and uncertainties to this basic picture. What sets off the process? There are many ways to do it, and the more complex an organism, the richer the triggering repertoire of its constituent cells. The initial signal can come from the outside, or from within the cell itself. Many external 'death signals' are proteins that bind to 'death receptors' on the target cell's surface. Cells can also trigger apoptosis themselves, for example if mitochondria start to leak ions because of oxidative stress, if the cell's DNA has been damaged, or if some other

vital indicator is going south. Our picture of apoptosis biochemistry gets more complex by the week
and so does the relevant literature, which keeps
increasing exponentially. But all of this bewildering
information is held together by beautiful genetic
experiments in worms, fruit flies and mice, which
define the order of steps, their relevance, and sometimes their redundancy. A growing male of the worm
Cenorhabditis elegans forms a total of 1179 body cells,
of which 148 are condemned to die by apoptosis
before the worm has fully matured. For the her-
maphrodites, the corresponding numbers are 1090
and 131. This simple worm sets aside no fewer than
13 genes for the control of apoptosis and all but two
of these genes function within every body cell.

How many genes control apoptosis in humans?
The latest official tally has broken the one hundred
barrier, and I would not be surprised if the final number turned out to be several times higher. These genes
control the modification of proteins by the attach-
ment or removal of phosphate groups or by short-
ening the protein chain, the movement of proteins
within the cell, wheels within wheels, you name it.
The clockworks for orderly death and orderly growth
are very much alike. The resemblance is more than
coincidental, because apoptosis plays a pivotal role in

the development and growth of complex organisms. When I was still in my mother's womb, it sculpted the fingers on my emerging hands; later on, it helped set up the wiring of my brain; still later, it delivered me from immune cells that would have turned against me; and all along the way, it purged my body of cells whose growth clock had gone awry and that were threatening me with cancer. I hope that programmed death will continue to look after me. May it be alive and kicking until the end of my days!

Unlike chaotic death, programmed death can be retarded or even blocked by specific mutations. Such mutations may inactivate an essential death gene or activate a death-inhibiting gene. When our physicians try to kill cancer cells, they usually resort to poison or radiation and inflict grave collateral damage. If they only knew the secret password for triggering apoptosis in cancer cells, they could simply ask these cells to please leave. I expect to see a host of new drugs that save human life by triggering cellular suicide. Let's hope such drugs will be around when I need them.

Programmed cell death raises many profound questions; philosophers should fall all over themselves staking out this intellectual gold mine. I am intrigued by the fact that full-fledged apoptosis only appeared with the advent of complex organisms made up of different cell types. As long as 'life' was synonymous with 'single cell', there was no need for apoptosis. What mattered was growth. It was life's nomadic era in which it was every cell for itself, and in which the hero's laurels went to those that could overwhelm their neighbors by sheer numbers.

But as cells amassed ever more genetic information, got used to the luxury of differentiation, and assembled into ever more complex organisms, the cellular *laissez faire* of the olden days became dangerous. Cells still had to grow, of course, but now they also had to know when and where to stop or even disappear in order to make room for others. Life's new civilized era reserved the hero's laurels for the wise rather than the brave. Smart cells with their large genomes did not talk more, but knew how to say the right thing – or to say nothing – at the right time. Differentiation and embryonic development required the discipline to keep silent and the art of stepping back.

The pitiless laws of evolution should never guide our own behavior. If they did, humans would stop being humane. Evolution cannot teach us ethics, but it can tell us much about running a complex enterprise. Evolution's business is living matter, the most complex form of matter we know, and its business experience is some 4000 million years. Should you come across a more impressive professional résumé, give me a call. If it comes to practical advice on how to handle intricate systems, nothing beats evolution.

Reflecting on apoptosis can give us such practical advice. Most of us are no longer nomads, but we still cling to outmoded nomadic ideals. We adore the hero who goes it alone, be it with a sword, a six-shooter, or a laser gun. Such heroism will always have its place, but our fixation on it destabilizes our complex and differentiated societies. We should give the hero's laurels to the quiet heroism of *médecins sans frontières*, social workers, couples raising orphans, or altruistic political leaders. We should give it to those who can step back.

Knowing when to step back is also the hallmark of understanding parents and good teachers. Many of them try to shape their children and students into a preconceived mold instead of letting them find their own way. Domineering parents and teachers have damaged more young people than overly permissive ones. Faculty members who constantly interrupt a student's seminar make me squirm, because watching over a talent calls for respect and patience — and the inner strength to keep silent at the right time.

Even science administrators might heed this advice. Managing science demands intelligence, organizational acumen, and scientific intuition, but also respect for the mystery of human creativity. Creativity is a delicate flower that wilts quickly when manipulated. Science policies that restrict research to 'relevant' questions and that tell researchers how to do their research suffocate creativity and innovation. They forget that the more specific the question, the less surprising the answer.

My chemistry studies have earned me a doctorate in philosophy – I am officially a 'Dr. phil.' The 'phil.' strikes some as a little quaint, yet I am very proud of this little suffix. I took up chemistry because it promised to explain many things about myself and the world around me, and chemistry has kept its word. But soon after my doctorate I stumbled across Erwin Schrödinger's little book *What is life?* and was hooked. It seemed like the ultimate chemical question. I went into biology because I expected to learn about the mechanism of life. Now I also have learned about that of death. It comforts me that the two are so similar. Life on our planet is so strong because it has mastered the art of stepping back.

NETWORKS,
FRETWORKS

IN MARCH 2000 THE SCIENCE MINISTERS OF THE
European Union (EU) met in Lisbon and declared
that within ten years the EU should be 'the world's
most competitive and dynamic knowledge-based economy.' Yeah, sure – and in 1893 US law makers pro-
claimed that 'the metric system should be the country's
fundamental standard of length and mass.' I don't know
what you say to such proclamations, but I know what
my Uncle Paul would have said. A peasant's son who
had made it to head waiter at a posh New York City
club, he could spot bull through a brick wall. 'Fat
chance,' he would have said 'sounds to me just like dem

toisty wise guys at the club.' Vintage Brooklyn, Uncle
Paul was.

He was also usually right. It is indeed unlikely that
by 2010 our 'European Research Area' will have outdis-
tanced the USA in science and technology. If anything,
we seem to be falling further behind. Yet we have so
much going for us. We have the same living standard as
the USA, about the same per capita income, and perhaps better general education. We even publish more scientific papers than the USA. In spite of all this, I would
guess that at least two thirds of all truly fundamental
biomedical innovations come from the USA. Why?

Some of the reasons are well known: antiquated
university structures, exclusion of women, and lack of
money. But Europe also suffers from serious misconceptions about how science works and what it needs.
Many of our research programs implicitly assume that
fundamental discoveries can be planned; that research
is more efficient if the scientists are told what to do,
and how to do it; and that coordination, cooperation
and evaluation are substitutes for individual talent.

Because we scientists have not done enough to
fight these misconceptions, they are contaminating
our science establishments and strangling scientific
innovation. The general public sees science as a logical, organized and even pedantic exercise in which the

scientists patiently put stone upon stone until the meticulously planned building is finished. Yet good science is exactly the opposite: it rarely follows a planned course, but is intuitive, full of surprises, and sometimes even chaotic. Scientific exploration is an expedition into the unknown – just like good art. The greater a scientific or artistic innovation, the more it surprises us. How could anyone expect to plan it?

If you want to know Europe's answer to this question, read the instructions for applying to one of its official research programs. Many of these intend to promote basic research in an area which happens to be fashionable: neurobiology, cancer, information processing, gender studies, genomics – or *nano*-anything. Many of them stipulate that you must collaborate with other research groups – you must form a 'network'. Usually you must also make sure that the members of your network represent a politically correct balance of geography and gender.

You are encouraged to budget a part of the funds for administrative purposes, and for good reason: once the network is underway, it will generate mountains of paper. The application is a harbinger of things to come. It must give all the details on every participant and may easily top a hundred pages. Everyone must describe in great detail what will be done during the next three or even five years, even though everyone hopes that unexpected discoveries will soon make the research plan obsolete. Some application forms even ask for 'milestones' – a timetable for the discoveries to come. Milestones may facilitate short-term technology development for which the road has already been mapped, because the basic knowledge is available. But how can one demand milestones for an unknown road? In fundamental long-term research, milestones are ridiculous.

In spite of all these hurdles, scientists will play along with such official network programs because they cannot afford to ignore the carrot dangling in front of their nose. Of course, most of them will not really change their research to fit the bill, but will only custom-tailor their application. If they are biologists, they will promise to combat ageing, prevent Alzheimer's disease, or cure cancer. But bending the truth undermines public trust in science and the scientists. Also, it makes scientists become cynical. A radio journalist recently interviewed one of my friends who had just landed a network grant worth several million Euros. My friend is an eminent scientist and as critical of enforced networks as I am, yet he praised the program into the sky. 'What else could I have done?' he apologized to me later, 'should I have said that giving me all that money was a bad idea?'

For young scientists trying to establish their own identity, official networks are rarely helpful. Young researchers want to show what they can do on their own, and may not always be eager to share their latest ideas with more established colleagues. When twenty-two top young biologists were recently asked whether they thought that their country's official network program helped them in their career, twenty-one of them answered 'no'. Network research also makes serious quality control very difficult, because a large group of scientists will nearly always produce some results, and because the evaluations rarely ask the right question. They usually assess a network's accomplishments, yet the real question is whether the funds would have been more effective had they been given to individuals without any strings attached. Because this second question is virtually impossible to answer, most network evaluations are bound to be inconclusive — they are experiments without a valid control.

The folly of such mandatory network schemes struck me forcefully last year, when helping a foreign country evaluate applications for a Genomics Network Program. I have always enjoyed serving on grant panels or prize committees, because they give me the chance to steer money and prestige towards scientists who deserve them, and away from those that don't.

Spending a day on a good committee always makes me feel useful, even virtuous. But that long day on the network panel left most of us on the committee depressed. We knew the applicants personally and were pained to see how they tried to dupe us by posing as genomics experts. We were prepared to overlook these antics, but had a hard time to rank the applications. Is an application submitted by six average scientists better or worse than one by four bad and two excellent ones? I opted for the latter, but not everyone shared my view. In the end, we struck out the weak parts of some applications and tried to strengthen the good parts by recommending additional studies, or the recruitment of other scientists. Instead of judging the applications, we were rewriting them! To cap it all, we had to reject the only application we were all excited about, because it did not meet the narrow specifications of the network program. We ended up doing a bad job, because we were not allowed to do a good one. Everything was upside down.

Telling creative minds what to work on, or how to do it, is a sure way to prevent exciting new things from happening – in science as well as art. Most of us would agree that the best way to hear good music is to pick good musicians. Why not follow this sim-

ple recipe also in science? Here is the answer given by the position paper of October 4, 2000 on the European Research Space: 'European research efforts should be focused on a more limited number of priorities which should be subject to a political choice on the basis of objective assessment criteria.' In a 1941 interview for the British science journal *Nature*, the Soviet Ambassador to Great Britain, Michailovich Maisky said it more succinctly: 'In the Soviet Union there is no place for pure science.'

There is nothing inherently wrong with scientific networks. On the contrary, they can be powerful instruments of scientific innovation. Science is intensely social and thrives on contacts and collaborations. The World Wide Web, the network par excellence, is the brainchild of scientists who wanted to stay in close touch with their colleagues around the globe. Another classic is the spontaneous collaboration between James Watson and Francis Crick on the structure of DNA. Even top-down networks can be useful if they focus on technology development and clinical research, where many different approaches must be directed towards a clearly defined objective. But in fundamental research, networks should only form spontaneously and when needed. Trying to enforce them is against the way scientists think and work. Scientists love to *cast* nets, not to be caught in them.

How do scientists get their ideas? What makes some of them – as Albert Szent-György once remarked – 'see what everyone sees, and think what nobody has thought before.'? I do not know, but the mystery of the process has always awed me. Perhaps some of us have retained the gift of a child's playful ways – making up crazy words, or putting a hat on backwards. Perhaps it takes this child-like freshness to see that the way from A to C does not lead through B, but through X or W. Scientific and artistic creativity issue from the same mysterious springs deep down in us, which dry up quickly when we make them into a communal water supply. That's why truly fundamental discoveries nearly always come from talented individuals, and not from organized groups. If we want new ideas, we should select the best brains, give them what they need, and then let them find their own way. This simple policy does not reflect submission to scientific arrogance, but respect for the vulnerability of human creativity.

Enforced networks are only a symptom of a more fundamental problem – Europe's growing mania to micromanage innovation. The root of this problem is the fact that Europe's best scientists do not shape Europe's research policies. Good scientists instinctively know that an effective science policy must be willing to step back and let creative minds find their own way – even if the way appears unconventional and risky. They know that there is no innovation without mistakes. If evolution had shunned mistakes, we would still be bacteria.

Typical administrators, however, are trained to avoid mistakes, because it is their *raison d'être* to execute decisions from higher up, smoothly and according to regulations. If administrators are not given clear guidelines, they take over by default and run things their way. They will leave nothing to chance and, with the best of intentions, over-regulate and inhibit innovation. That's what has happened to most of our EU science programs. They are being strangled by a Byzantine bureaucracy, because so many of our best scientists have turned their back on science politics. They prefer to stay at the bench or in the library, and relegate science politics to second-rate scientists or to administrators. Listen to this advice: 'It is essential for men of science to take an interest in the administration of their own affairs or else the professional civil servant will step in – and then the Lord help you.' That's the great physicist Lord Ernest Rutherford speaking – more than three quarters of a century ago. Or, as Uncle Paul would have said: 'You gotta do it yisself, or it ain't gonna woik!'

EURO-BLUES

I OFTEN THINK OF JAN. HE HAD ALWAYS wanted to be a biologist and started out by doing a PhD thesis in microbiology at a European university. His postdoctoral work in the USA must have gone well, because his American hosts invited him to apply for a faculty position. But Jan wanted to try his luck back in Europe and took a job as 'assistant' at a university in his home country. As this position was on soft money and limited to two years, Jan soon switched to a six-year assistantship in another department. Now this position is about to run out and Jan is in a bind. The next steps would be either an 'habilitation' or an official 'teaching assignment', neither of which would offer him a permanent position or the promise of a professorship. Like so many scientists his age, Jan is lost in the maze of a fuzzy academic career structure. He often wonders whether he should cross the North Atlantic once again, but this time with a one-way ticket.

There are three significant aspects to this story. First, it is true – except for Jan's name and perhaps his gender. Second, it is not only about an individual, but also about an entire scientific generation. And third, Jan is European – Swiss, to be exact. If he were French, Belgian, German, Spanish, or whatever, his position would have a different name and different problems, but it would still be part of the Mediocre European Science System, appropriately acronymed MESS.

Europe is busy overhauling its creaking universities. The reforms are generally pushed by governments bent on saving money, and smash head-on into academic inertia or outright resistance: the irresistible force meeting the immovable object. Most of the reforms focus on politics, finances, and academic governance. Only rarely do they pay attention to the young academics. And if they do, young academics rank way down the list.

Yet a university is first and foremost about people. When will our European policy makers wake up to the fact that a uniform, selective and fair academic career structure is the most pressing problem of European science?

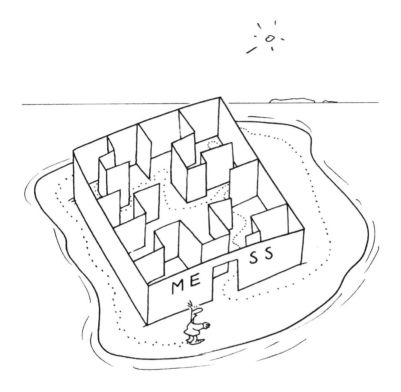

Top research needs top people. The same goes for teaching, which at a university should always be tied to research. Excellence in research and teaching needs above all creativity. But creativity is not a commodity we can produce at will. We can (and often do) suffocate it in any number of ways, but cannot make it jump out of a box whenever we need it. We cannot simply tell it to happen. It is a gift that each new generation gives us in the form of new talent. This talent is our most precious resource. We should prospect for it, mine it diligently, and be careful not to waste it.

It is the young who do most of the research at our universities. Older academics have many other duties and usually cannot spend enough time in the laboratory, the library, or the field. They might set the research goals, discuss results and publications, and run their research team. But the thrill of discovery, the eureka moment, usually smiles on the young. Not only because they have more time, but also because they are generally more creative than their older colleagues. There are, of course, many notable exceptions, but in general scientists do their most original work early in their career. We should nurture this phase by letting our young scientists follow their own ideas when they are particularly good at having them. We should offer them a career structure that is selective, transparent and fair.

Very few European universities offer such a career structure. Returning to Europe after a postdoc overseas can be a descent into a Kafkaesque netherworld of ill-defined positions, fickle judges, and scientific serfdom. Most early positions are for a limited term and cannot be extended. When they run out, the scientist may be left out in the cold. Permanent assistantships are still worse, because they trap young researchers in a permanent intellectual dependence. Even the last stage in this academic hurdle race can be touch and go. Vacant professorships are often filled within the same narrow specialty through murky selection processes, which hand the job to a former associate of the retiree.

Europe pays a high price for taking such poor care of its young scientists. One price is lower scientific innovation. If scientists cannot follow their own ideas while they are most creative, they will make fewer discoveries. Another price is lower technological innovation. Putting scientific knowledge to practical use calls for independence, motivation, entrepreneurial vision, and willingness to take risks. It takes guts. One does not develop guts by sitting in lectures or doing experiments on a professor's orders. One develops guts by emulating those who have them. Nothing is more important for developing a student's personality than admired role models. Students interact mostly with young assistants. If these have been postdoctoral fellows abroad and now must serve their professor's research, they will be frustrated and hardly inspiring examples of courage and creativity. Independence and willingness to take risks are not only the essence of technological innovation, but also of political and administrative innovation. Yet these are the very qualities our academic system selects against.

Some Europeans counter such criticism by pointing to their country's excellent scientific publication record. But this argument misses the point on several counts.

Some rich countries have papered over their antiquated academic career structure by funding research generously and hiring established foreign stars. Such a policy may be good for science, but does little for young scientists. It also deprives our students of young role models. And it makes Europe lose the international competition for young scientific talent. If some of our best young scientists leave Europe for the United States, so be it, as long as the opposite is also true. But young stars from top academic institutions overseas rarely accept junior positions at European universities, because these offer such unattractive career prospects. Instead of exchanging young talent, we export it.

Should Jan decide to become an assistant professor in North America, he would face a system that differs from most European ones in three important ways. First, the assistant professorship is advertised internationally, and applicants are screened as rigorously as those for a full professorship. Second, the successful candidate receives a written five- to six-year contract that guarantees the following: independence in applying for research funds and running a research group; full participation in almost all departmental decisions; and only limited teaching duties. Third, one year before the contract expires, research, teaching and academic citizenship are evaluated with the help of experts from around the world. If the result of the evaluation is negative, the assistant professor must leave the university within one year. If it is positive, the assistant professor is promoted to a permanent professorship without competition with others. It's 'up or out'.

This *tenure track system* is tough, yet transparent and fair. Universities in the USA and elsewhere have used it for decades with great success. Like any successful product, though, it constantly fights against shoddy imitations. And Europe leads the world in concocting counterfeit tenure tracks. Some of them grant independence, but do not reward success by a promotion. Others burden the young academics with a full teaching load and unrealistically short time limits. Still others call for rigorous selection, but offer permanence right from the start. I sometimes wonder who cooks up these bizarre schemes.

There is a serious catch, however: tenure track only works if the university works. Tenure track is like a CAT scan of a university's vital functions. For example, it tells us whether the university has a strong and long-term academic leadership that can plan vacant positions in advance, because for each assistant professor hired today, there must be a vacant permanent position five to six years down the road. It also tells us whether the university has an efficient selection system, because choosing a young and still little known researcher is much trickier than wooing an international star. On these counts, very few European universities emerge with a clean bill of health. Their Faculties, God bless them all, are usually too heterogeneous, too inefficient, and too political, and single university institutes are too vulnerable to tampering by powerful individuals. Promotion of assistant professors is best handled by well-run, large departments or long-term, independent deans, whose decisions must pass scrutiny by a separate and independent presidential *ad hoc* committee. Tenure track also demands that mandatory retirement function properly. If permanent professors refuse to retire, as now often

happens in the USA, the system becomes unstable because influx and outflux no longer match.

Tenure track does not imply that all permanent professorships are filled through tenure track. Launching a new institute or research initiative, or focusing existing strengths, may call for the hiring of an internationally-known researcher as a permanent full professor. Tenure track may also be unsuited for some 'small', but important disciplines such as archeology, because these offer so few full professorships. These 'small' fields, as well as disciplines that must teach very large classes, may have to hire some of their assistant professors without tenure track. While such a dichotomy is not ideal, it is at least transparent because candidates would know from the start what they were getting into. A wise university leadership will always aim for a healthy mixture of tenured and untenured faculty.

Where does this leave the other early academic career steps in continental Europe, such as the *maître-assistant(e)*, the *ayudante*, the *maître de conference*, the *hoofd-docent*, the *Oberassistent*, the *ricercatore*, and the *Privatdozent*? Exactly nowhere. Each of these positions has something seriously wrong with it. Each of them is a long and uncertain voyage without a clear path to a permanent professorship. They do not offer a track to a tenured position, but a trek through a penured one.

Tenure track is no panacea for the many ills of European universities. But it would encourage and reward excellence, and should go a long way in helping young academics to move within Europe. Europe should adopt this system now. If Jan were to leave Europe for North America, he should do so in a spirit of curiosity and adventure, and not because Europe denies him the chances he deserves.

Having grown up to the wail of air raid sirens, Europe's unification has been one of the most inspiring events of my life. I wish I could say the same of Brussels' science policies. They fiddle too much with science, they are too political, and they do not pay enough tribute to excellence and our young scientific talent. They do not think enough of Jan. Europe's scientific prospects tickle me pink and should make others green with envy. But the reality just gives me the blues.

13

THE SEVERED CHAINS

W HAT MAKES ME HUMAN? I ALWAYS thought that my distant ancestors had picked up some genes that severed the chains tying them to their ape-like relatives. Maybe so, but genome-gazers have not yet found these liberating genes. Perhaps I looked at the problem the wrong way around. Perhaps I am human because my ancestors got rid of genes that kept them from becoming human. Perhaps these ancestors were smart enough to know that sometimes one must lose genes to gain freedom.

Human behavior is complex and often unpredictable. At least mine is, if you believe my former students and postdocs. I never took their complaints personally. Why should I? It is my sensory receptors and signal processing circuits that are responsible for the chemical communications within me, tell me what is going on outside, help me think about it, and then shape my moods and decisions.

All of my sensory receptors are proteins and a few of them also have a small organic molecule attached to them. Most of them are firmly stitched into the membrane that surrounds each of my cells, meandering back and forth across that membrane seven times. They are my 'seven-transmembrane receptors'. How many different ones do I have? I do not know for sure, because the amino acids in their protein chains are arranged in so many different ways that there is no sure way to identify all the receptor genes in my genome. Some of my receptors form subgroups whose members have a similar sequence of amino acids, but overall my seven-transmembrane receptors share only their seven transmembrane spans and their role as antennae for signals from the cells' outside. I have at least a thousand different seven transmembrane receptors, but there could be hundreds more. Between three and four percent of my genes are set aside for them – by far the largest gene family in my genome. Three to four percent may seem extravagant, but most of the families I know spend at least as much of their budget on communication – and that's not counting those with phone-addicted teenagers.

All my seven-transmembrane receptors work by the same basic mechanism. First, they bind the incoming signal in a specially shaped pocket formed by their intricately folded protein chain and then change their shape. Second, in this new shape they attract an intracellular protein (nicknamed 'G-protein') that has three different polypeptide chains and a tightly bound molecule that contains a sugar, an organic base, and two phosphate groups. Biochemists call such a molecule a 'nucleotide', and refer to this particular one as GDP. Third, the G-protein exchanges its bound nucleotide for another one (termed GTP) that differs from GDP by one additional phosphate group. Fourth, the GTP-charged G-protein activates an enzyme that converts yet another nucleotide (termed AMP) to a cyclic variant (termed cyclic AMP). Finally, cyclic AMP can either open or close an ion gate in the cell membrane and thereby change the membrane's electric potential, or activate or inhibit other enzymes inside the cell. Although there are many variations to this signal processing circuit, all of them amplify the original signal by as much as a million-fold or more. And amplification can continue in the target organs, such as the muscles – just think of a six-ton elephant scared into a run by a few photons hitting its retina.

My seven-transmembrane receptors are tuned to many different signals. Some are tuned to proteins, peptides or small organic molecules in my body fluids – they are hormone receptors. Four of them – the different rhodopsins of my retina – are tuned to light and color. But the vast majority of these receptors keep track of smells and tastes.

I have about 900 genes for different smell receptors, but more than 60 percent of these genes are defective in some way – they are pseudogenes. That leaves me with about 400 intact smell receptors – still nothing to sneeze at. But a mouse has three times as many and a rat even more. Most of these receptors sit in the smell-sensitive nerve cells of my nose and register the hundreds of thousands, perhaps even millions of different chemicals in the air I breathe. Each smell-sensitive nerve cell harbors only a single type of smell receptor and feeds its signal to the 'olfactory bulb', a small organ within my brain that compares the signal from one receptor with those from the others and then feeds the processed information to my forebrain. That's why I can detect subtle shadings of fragrances, such as the difference between a Bordeaux and a Burgundy. To a wine lover, that's hardly a big deal, but I am proud of it.

My sense of smell is important for me, but I could manage without it because I can also see and hear. This may explain why we humans accumulated defective smell genes much faster during evolution than all other known animals. Worms such as *Cenorhabditis elegans* badly need their sense of smell, because they are deaf and blind. *Cenorhabditis elegans* reserves about 5 percent of its genes for smell receptors. That's a heavy investment, because the worm needs almost as many genes for basic cellular functions as I do, yet has a genome 30 times smaller than mine.

I can not only *smell* chemicals with my nose, but also *taste* them through my tongue and palate. I can distinguish five tastes: bitter, sweet, *umami* (the taste of monosodium glutamate), sour, and salty. I sense the first three of these tastes through 'seven-transmembrane taste receptors', and the other two through other mechanisms. I do not know how many seven-transmembrane taste receptors I have; chemosensory heavyweights such as mice have about 40 of them, so I might have ten or twenty. Even though I cannot sense any shadings of sweet, bitter

or *umami*, taste enhances my sensory repertoire, allowing me to detect some ten thousand different aromas.
I cannot categorize them all; even professional 'noses'
who make a living sniffing perfumes, wines or spirits
can define at best a few thousand aromas.

But there is more. Many animals release chemicals
that induce an involuntary stereotyped response in other
members of the same species. Biologists call these
chemicals *pheromones.* Pheromones can travel over large
distances through air or water and control social status, mate selection, aggressiveness and hormonal status.
They are irresistible chemical commands that higher
animals may not even consciously perceive as smells.
Queen bees release a pheromone that prevents worker
bees from rearing another queen. Worker bees use alarm
pheromones to persuade their nest mates to sting an
intruder. Female mice choose Pheromone Speak to
inform interested males of their menstrual status. And
the pheromone emitted by a newly hatched female moth
can attract dozens or hundreds of eager males who will
travel for a mile or more to pay their respect. With
insects, pheromones are often the only means by which
individuals of one species can find one another.

Pheromones come in a huge variety of chemical
structures and usually function as chemical mixtures,
the exact composition of which is crucial. If one

component of the mixture is missing or present in the wrong proportion, pheromone function may be lost, elicit the opposite effect, or affect the wrong species. In higher animals, pheromones may be steroids or proteins, and trigger discrete sensory nerve cells in a sub-region of the nose, the 'vomeronasal organ'. These 'vomeronasal receptor cells' work with a separate family of seven-transmembrane receptors and distinct downstream signal processing circuits that feed into a special processing center, the 'accessory olfactory bulb'. This bulb sends the processed signals not to the brain cortex, the site of voluntary decisions, but directly to the limbic system that controls involuntary hormone secretion and behavior. A mouse has about 300 different vomeronasal receptors, most of them probably tuned to pheromones.

Even we humans, in spite of our complex and malleable brains, may be unwitting slaves to pheromones. Women living together in college dormitories or Bedouin tents generally menstruate in synchrony, because they release substances that regulate the menstrual cycle of others. These substances can be collected from the armpits and are not perceived as odors, suggesting that they are pheromones. There are also hints that smells linked to certain sugar-containing proteins (glycoproteins) may influence our sexual attractiveness and aggression towards others. These glycoproteins are encoded by a large and complex region of the mammalian genome, which biologists refer to as the 'MHC1 locus'. During sexual reproduction, the hundred or so genes present within this genomic region can recombine in various ways, creating up to 3600 million different new genes which in turn can direct the synthesis of as many different *MHC1* glycoproteins. As the world population now stands around 6000 million, each of us (who is not an identical twin) has, on the average, less than one *MHC1 Doppelgänger* somewhere on this planet.

We normally reject tissue grafts from other individuals because their *MHC1* proteins do not match ours. But *MHC1* proteins can do more than trigger rejection of foreign tissue grafts. Pieces of them can escape into the blood stream, the sweat and the urine, and contribute to individual body odor. In mice these odors clearly determine the selection of mating partners and aggression towards other mice. Female mice prefer males whose *MHC1* proteins differ from their own, and the sketchy experiments with humans point in the same direction. It boggles the mind: the same molecules govern recognition of 'non-self' by single cells, and complex psychological interactions between human beings! It is an ode to the parsimonious design of life on earth. And it is also a story that sends a shiver down my spine.

Controlled human experiments on sexual attractiveness are difficult and touchy. They also fascinate perfume makers, who flood the World Wide Web with wacko reports on pheromone-based 'irresistible sex attractants'. So it is no surprise that the effects of secreted *MHC1* proteins on human behavior are still poorly documented and highly controversial. It is also open whether these glycoproteins are typical pheromones, and whether they work directly or through indirect mechanisms – for example by promoting growth of odor-releasing bacteria.

Obviously, pheromone-controlled mate selection could help mice – and humans – to maintain genetic diversity. That's fine with me, but how adamant is this chemical directive? Why had I asked that particular girl to be my dancing partner for our high school's senior prom? I hate to think that it was just her *MHC1* proteins. There may well have been other chemical signals that shaped my decision at that time. Indeed, I consider it very likely that we shall uncover additional human pheromones, and that not all of them will have to do with attraction. There must be a deeper reason why the French and Germans, among others, refer to an unpleasant type as someone 'they cannot smell'. Does our liberal use of deodorants throw a monkey wrench into this delicately balanced communication system? I am convinced that there is a molecular biology of hate and love even for us humans, and that we should learn more about it. But there are moments when I would prefer not to think of it.

Human pheromones frighten me, because they are a potential threat to my humanity. If I want to find out who I am, there is no way around the question of how much I am tyrannized by pheromones. To my relief, the probable answer is 'not very much'. I lack an accessory olfactory bulb and a typical vomeronasal organ, even though I may have had vestiges of it at birth. Also, more than 95 percent of my putative pheromone receptor genes are dead pseudogenes and even the few intact ones I have may be useless because I lack key parts of the downstream signal processing circuit. I cannot exclude that some pheromones work through my smell receptors, but overall my pheromone-related genes are a colossal genetic junk-yard littered with evolutionary debris. That's very, very good news. Here are the pieces of the chains that had kept our ancestors in chemical bondage. To become humans, they had to invent genetic wings that let them soar. But then they had to sever the chains that tied them to the ground.

FIVE EASY STEPS TO
GET RID OF YOUR
LAB

S O YOU ARE NOW A PERMANENT PROFESSOR —
a 'Certified Academic'®. Congratulations! But
don't get carried away and expect the academic
River of no Return to wash you gently into the Bay
of Retirement. In fact, you have just been handed a
life sentence of forced hard labor. Setting up your
first lab as an assistant professor was your academic
honeymoon, but now it's time to face reality. Those
problems you overlooked at the start will grow on
you and pester you for the rest of your career. As the

head of a lab, you will always have to toil for others. Year after year, every student and postdoc will bother you with umpteen problems; settling the authorship of papers will require the craftiness of a lawyer, the patience of an Inuit trapper, and the impassiveness of a Buddhist monk. And trying to rake in the funds for keeping your lab humming will chain you to a computer for most of the time. It will be like putting a new coat of paint on the Golden Gate Bridge – as soon as you have worked your way to one end, it's time to start at the other one. If you look at it squarely, you are done for.

But all is not lost. You have already shown the world – and yourself – that you can run a lab, so why not ditch it now and move on to bigger things? I am not talking about becoming a chairman, a dean, or a university president – they all serve their own harsh sentences. I am talking about Easy Street – learned societies, think tanks, obscure scientific academies and advisory bodies, where almost everyone is either a president, a Secretary General, or on an expense account. You, too, can do it – just about everybody can. Only that lab of yours stands in the way. You cannot simply walk away from it or return it to the sender, but you can make it go away all by itself. It only takes a little patience – and five easy steps.

Step 1: Accept as many students and postdocs as you can. Never ask the others in the lab what they think about the people who apply to you, and don't worry about whether there is enough space. Be vague when you assign individual bench spaces or, better yet, make several people use the same bench space *on a rotating basis.* And while you are at it, put several of them on the same research topic and keep consumables in short supply. Having your students and postdocs step on each other's toes and bump against each other's brain is a great way to keep adrenalin levels high and general happiness low.

Step 2: Never enter your lab. Have your people trudge to your office, which should be as far from the lab as possible and reflect your exalted position. If necessary, use creative bookkeeping to convert grant money into plush carpeting, designer curtains, an impressive executive desk for yourself, and uncomfortable chairs for your visitors. Everything about your office should convey the message 'You are speaking to the boss.' Don't waste time discussing unsuccessful experiments – you want *Publishable Results.* Insist on office hours (don't be too generous there) and keep looking at your watch to remind your students and postdocs that you have more important things to do than to talk to them. Always keep the door to your office shut, and put visitors in their place with blinking 'WAIT' and 'ENTER' signs. Above all, hire a fiercely loyal secretary who guards the entrance to your inner sanctum with the charm of an underfed pit bull. The office of a Very Important Person is usually empty, so don't stick around too much. Never turn down an invitation to join a committee, write a review or a book, chair a panel, consult for a company, or attend a meeting – especially a foreign one. A scientist of your stature should not sit in an office, but be either abroad, jet-lagged, or dashing to the airport.

Step 3: Pick a favorite among your group (let's call him Jim) and let everyone know. Remarks such as 'Why can't you be like Jim?' or 'Funny, Jim had no problems at all with this experiment!' are a great way to dent egos, make Jim a pariah, and divide your lab into warring factions. To keep them warring, complain to Jim about others in the lab. If you ask Jim to keep your complaints to himself, he will leak them verbatim to the rest of the lab as soon as you are out of earshot. If you want to go all out and can rise to the challenge, use the next lab party to start an affair with a member of your group. That's bound to be a bombshell and will guarantee that your lab will talk about nothing else.

Step 4: Never miss a chance to put your students and postdocs in their place. Refer to them as 'data grinders' or 'bodies', preferably when they are around to hear it. Interrupt them when they give seminars or progress reports until you have reduced them to stuttering wrecks. Keep them well separated from all seminar visitors – these are, after all, *your* guests, not theirs. When they ask you to write a letter for them or look at the draft of a manuscript, let them cool their heels for a few weeks before you respond – that will remind them that you are *real busy*. And when you talk about your group's work at international meetings, never mention your co-workers by name and let the audience know that you had 'burned up a bunch of postdocs on this problem until you yourself stepped in and solved it.'

Step 5: Don't be squeamish about wielding power. Trying to get rid of your students and postdocs will not make you their darling, so don't worry about what they say about you when you are not around – which should be most of the time. During the rare moments when you see them face to face and must suffer their flak, remind yourself that you will outgun them later many times over. Every single one of them will need letters of recommendation from you, and that will be your *jour de gloire*. You could refuse to write these letters, but why in the world would you blow your chances to get even and to show off your mastery of academic *double entendre*, which conveys the opposite of what is written? The epithets 'good', 'conscientious', or 'diligent' are poisonous arrows that will do their jobs and prevent the subjects of your letter from getting theirs. So will the remark that 'he (or she) might well mature further.' But if you really want to do somebody in, then use the term 'solid'. It means 'dumb' in plain English and will stop anybody from reading any further. In the code of academic recommendation letters, 'solid' is the stop codon. Recommendation letters are supposed to be confidential, but you can be sure that those for whom you wrote them will soon know every dot and comma of them and vent their feelings about you with gusto for the rest of their life.

Those five steps should do it. As you can see, curing you of your lab is not nearly as hard as it seems. And as long as the academic grapevine stays as lively as it is, the cure will be permanent. What more can you ask for? On the strength of your 'past research experience', you will be free to roam new pastures where the grass is greener, everybody's hair grayer, and the offices even more impressive. But it might be a good idea to keep that secretary.

15

THE RISKS OF
PLAYING SAFE

WHEN I WAS SCIENCE ADVISOR TO THE Swiss government, my friend Walter sent me a bulky document that looked like the manuscript of a book. I wondered where Walter had found the time to write his opus, but then noticed with dismay that it was the application for a network grant, which Walter had put together for himself and a dozen other colleagues. Most of it was about finances, timetables, professional résumés, organizational details, and the usual statements from the institutional officials; less than half of it was about science. Walter, a stellar biologist and highly creative mind, must have spent untold hours putting together this monster. Why on earth would our National Science Foundation waste supercomputers on tallying grocery bills? Something was out of whack.

Walter's case is not unusual. Wherever I look, public systems for science funding are drifting off course. They should stimulate novel discoveries, but increasingly encourage short-term pedestrian research. If you apply for your first research grant and dare to venture into new territory, you probably won't get funded because you 'lack prior experience'. (Apparently you are supposed to continue what you did as a postdoc). If you do have the experience and propose to tackle an ambitious and risky problem, they will damn you for 'going on a fishing expedition' and advise you to be 'down to earth' – to do experiments that are bound to work.

And if you get funded, you will probably have to apply for renewal within less than three years. This means that you will have to beef up your Progress Report with results you already had before you applied for the first round – or you will be labeled 'unproductive'. And you will have to lie. You will have to promise practical benefits that you do not really believe in; you will have to present detailed research plans, timetables and sometimes even 'milestones' that violate everything you know about the uncertainty of innovative basic research. If you want to survive, you must play along with the subtle corruption of the system. Your price will be disillusionment and cynicism – and less scientific innovation. As you probably have your best ideas while you are young, your first years of independent research are particularly powerful engines for scientific innovation. Our way of funding science makes this engine stutter and also undermines your honesty and enthusiasm. It forces you to divert too much of your time from research to writing grant applications, particularly if you are still young and lack secretarial help and collaborators. To paraphrase Albert Einstein, our granting agencies have perfected the means while confusing the goals.

How did matters get that far? A major culprit is aversion to risk. We Europeans are probably world leaders in zero-risk mentality, but we no longer hold universal patent rights on it. This mentality rears its ugly head whenever organization or administration spins out of control. In trying to prevent waste of public funds, many of our research funding systems are obsessed with preventing failures, unexpected problems, surprises and exceptions. But failures, unexpected problems, surprises and exceptions are at the very heart of scientific research. Research is an expedition into the unknown — that's why it is so exciting. A funding system that over-regulates research in the name of efficiency saps creative potential and collective wealth. It wastes the very resources it should protect.

The urge to control and predict research can grow bizarre flowers. When completing the final report for one of my grants, the form sheet asked me 'Did you obtain the results you expected?' What cheek! I was tempted to answer 'Of course not!' but then thought better of it and resignedly typed in 'Yes'. It seemed like the wrong place to be the hero.

Other innovation systems are not as timid — look at the biotechnology sector. In spite of all its hype and flops, it is one of the most dynamic and innovative activities in today's life sciences. But it is not for the faint of heart, because the time from discovery to market is long and most start-up companies go belly-up within the first few years. Yet the system prospers because in the long run the few winners more than make up for the many losers. Venture capitalists know that investing in the early phase of a start-up company increases not only their potential pay-off, but also their risk. They take it for granted that profit and risk are Siamese twins. Nature knows it, too; the evolution of life was the ultimate gamble — and the gamble is far from over.

Every decision to hire a young scientist or to fund a research project is a calculated risk. But taking risks is not the same as being reckless. Anyone who takes risks without professional know-how, experience and good judgment will soon have to pay a steep price. Our present peer review system – in which a group of experts judges the risks and the potential of individuals or research projects – is essentially a device for scientific risk control. But this device has started to fail because it must make unreasonably stringent selections. When there is only money to fund one out of ten research projects, the 'no' of a single committee member is usually deadly and even a competent and fair-minded group of peers will hand down erratic judgments. Like any risk control device, peer review also selects against the exceptional. It is a great equalizer. It encourages blandness and selects against novel ideas that challenge accepted dogma.

The leveling influence of peer review is not limited to science, but also impoverishes the performing arts. Today's young singers or instrumentalists who aspire to international fame must win international competitions in which a jury of experts picks the winners. Peer review again. A brilliant young pianist confessed to me that in these competitions he never played the way he felt, because a highly individualis-

tic interpretation was bound to rub one of the jury members the wrong way. Peer review was forcing him to play up to a generic taste and to adopt a bland style that was least likely to displease. No wonder that so many concert performances around the world have become stereotyped. If creative activity is subjected to Darwinian selection by groups of experts, the result is often timidity, standardization, and mediocrity.

A zero-risk mentality reflects lack of courage, the key ingredient of scientific success. Success in science depends on many factors — intelligence, perseverance, talent for leading and inspiring others, organizational skills — but none of them outranks courage. It takes courage to face the grueling selections of academia, to choose a difficult research problem, or to challenge an idea everybody accepts as dogma. If you want to discover a new source, you must dare to swim against the current. Because many of our research funding systems lack courage, they hide behind numbers and try to quantify the risk of a basic research project, or of a scientific career, as 'precisely' as possible. They are hooked on citation frequencies, impact factors, grant scores and university rankings. More and more, these phony numbers now decide hirings, promotions, and the flow of research funds. Their 'precision' satisfies everybody's longing for 'objectivity' and 'transparency', but nobody seems to care a hoot about the questionable methods by which these numbers are concocted. And once these numbers have escaped from Pandora's Box, nobody can put them back in.

University rankings are particularly obnoxious, because newspapers love them even more than administrators and politicians do. 'WORLDWIDE RANKING OF UNIVERSITIES PLACES UNIVERSITY A AS *NUMBER 1* AND UNIVERSITY B AS *NUMBER 2*'. It's all so neat – everyone can understand that. And if it rankles you that your own university is only Number 67, it feels good to know that the university nearby is only Number 69. But how can one possibly rank institutions as complex as universities? Adding up Nobel prizes, impact factors, and outside grant money may make some sense with the natural sciences, but what about the humanities? Their research grants are usually small and they tend to publish their best work as books for which impact factors and citation frequencies are not available. But it does not seem to matter – most rankings simply ignore the humanities. It always amazes me that the ranking gurus get away with it. In their intellectual universe, Friedrich Nietzsche, Baruch Spinoza, Bertrand Russell or Ludwig

Wittgenstein would just be invisible black holes. And is a university with twenty departments of average quality better or worse than one with eighteen bad and two truly outstanding ones? The choice cannot possibly be *objective*. But the ranking bandwagon keeps on rolling and is seriously distorting our universities. Many of these now try to attract professors who bring with them the most Brownie points for the next round of ranking. And the humanities are scrambling for the electronic limelight by publishing their work piecemeal in highly specialized journals. The result is predictable: the number of journals explodes and the importance of papers erodes. It may rhyme, but makes no reason.

Our science enterprise is caught in a seemingly insoluble dilemma. On the one hand, too many scientists are chasing after too little money, and deciding who should get what is becoming ever more crucial – and difficult. On the other hand, the unprecedented power of the decision machinery has made this machinery too complex, too costly, too conservative – and often too arrogant. There is no simple way out and it would be naïve to hope for 'The Solution'. Each country does – and should do – things a little differently, and forcing all funding systems to work the same way would do more harm than good.

Moreover, I do not see any credible alternatives to peer review of individuals, research projects, and institutions. For better or worse, we are stuck with this instrument, but we should use it more cautiously and with a keener awareness of what it can and cannot do.

Here are four simple suggestions. First, funding agencies should keep in mind that more organization and control do not necessarily mean better science, and that every page researchers must write or fill in cuts into scientific productivity. We biologists want to get the Nobel Prize for Physiology or Medicine, not that for Literature. Second, each country should make sure that its researchers can apply to several different funding agencies, because monopolies are as harmful for research funding as they are for the general economy. Third, evaluation committees should rely more on scientific expertise and intuition than on *objective* indicators, and should aim for a healthy balance between the true and tried, and the innovative and risky. Fourth, we scientists should rid ourselves of the arrogant notion that managing science is best left to the dim-witted. Unless we all roll up our sleeves and put more thought and effort into shaping funding policies, our fragile *SS Science Enterprise* will get crushed between the Scylla of dwindling funds and the Charybdis of computerized evaluations.

As an ever smaller fraction of excellent research grants can get funded, picking the winners is becoming ever more arbitrary – and ever more painful to both the judges and the judged. Perhaps we should crank up our courage and tackle this thorny problem by unconventional means. Why not select the best applications in the usual manner and then let a lottery decide? Far from being frivolous, a lottery would let finalists that did not make it save face and spare evaluation committees the agony of arbitrary decisions. During the Vietnam War, the USA used a lottery system to select recruits for military duty. Even though special prominence or lots of money could sometimes beat this system, it served its purpose surprisingly well.

Sitting on review panels, prize committees and advisory boards has usually taught me more about my colleagues on the committee than about those I was supposed to judge. The last committee I served on was truly superb and motivated by the best of intentions, yet in retrospect I realize that with time we started to overrate ourselves and became subconsciously arrogant. Because we respected and liked one another, we also slipped into the habit of avoiding disagreement. 'Where all men think alike, no one thinks very much' noted Walter Lippmann, and he may well have scribbled it during a committee meeting.

I no longer serve on evaluation committees, but if I did, I would urge all committee members to read some of the grossly erroneous judgments which artistic and scientific giants such as Robert Schumann, George Bernard Shaw, Johannes Brahms, Robert Virchow, Otto Warburg, or most official art critics, passed on the ideas or achievements of some of their contemporaries. My favorite example is how Ernest Rutherford, one of the greatest physicists of all time, dismissed the possibility of nuclear power generation as 'pure moonshine'. When peers peer into the future, they goof up nearly as often as the next guy. Indeed, there is no obvious reason why 'experts' should be especially good at judging human creativity. But as long as we cannot avoid judging others, let's do it with caution and a healthy dose of modesty.

16

CHAUVINISM
IN SCIENCE

LL AROUND THE WORLD, I AM TOLD, there are nondescript buildings from which some governments monitor the world's telephone conversations. I find it hard to believe that this can be done, but friends in the know assure me that today's computers can indeed dissect the global gibberish, pick out keywords presaging danger, and track their source.

My brain is not nearly as omnipresent, yet it, too, monitors the conversations around me. Certain words, or the way they are said, trigger an internal alarm that alerts me to someone to watch out for. Our individual alarm triggers reveal the blueprint of our soul. That's why I usually keep mine a secret. But for the purpose of this article I shall divulge one of them: chauvinists.

The term 'chauvinism' eternizes the inflated patriotism of Nicolas Chauvin of Rochefort, one of Napoleon's soldiers. True patriotism, however, is a far cry from chauvinism. A patriot loves his own country; a chauvinist hates everyone else's. Where there is chauvinism, racism is usually not far behind. These two are soul brothers who travel in pairs. Chauvinism is racism wearing a tuxedo.

My alarm circuits were installed by the Austrian Nazis whose schools I attended as a child. They taught me that the British were arrogant, the Americans dumb, the French sneaky, and the Gypsies smelly. The Jews, of course, were all of that — and then some. This liberal education was offered to me free of charge six days a week, rain or shine. By a child's osmosis I learned to recognize the Nazis' vocabulary, their cadence of speech, even the way they walked. But someone must have garbled some wires because the circuits that were supposed to make me a good Nazi did the opposite. Today, more than half a century later, I can still sniff out a fascist from fifty yards against the wind. I am the lucky product of a well-planned education gone awry.

When the Thousand-Year Empire was finally flushed down the drain, I was only nine years old and many years went by until I found out what had happened. When I did, I was anxious to give my brain a thorough spring-cleaning. But clean water was hard to come by, because most of the intellectual wells of my country were polluted. The humanities, literature, the visual arts, even music: they all bore the scars of Third Reich ideologies. But the natural sciences had resisted tampering. Their international flavor excited me and I hoped that they would help me escape from the intellectual doldrums of postwar Austria. In my youthful idealism I saw science as a white knight, a slayer of lies and prejudice. Science would allow me to work together with people from different cultures. I would become part of an intellectual web that spanned the globe far above the man-made turbulences of languages, religions, and nations. A life in science would save me from having to face chauvinism ever again.

My road from chemistry to molecular biology took me through many countries and gave my family a panoply of passports: my wife is Danish, my son American, one daughter Swiss, the other Austrian, and her husband Russian. A delightful mess, just the thing to make some of my childhood teachers turn in their grave. I have always tried to give my laboratory an international flavor and have insisted that my students and postdoctoral fellows talk to each other in English. One of them thought I was rather overdoing it and marked the door to her lab with the sign 'ONLY BAD. ENGLISH ALLOWED HERE'. I could not resist scribbling below 'STILL BETTER THAN ONLY GOOD GERMAN'.

I soon discovered, however, that chauvinism can infest even science. I saw little of this as a student and postdoctoral fellow, when doing experiments and getting a job were all that mattered. But as I started to work for international foundations, scientific organizations, or prize committees, my internal alarms went off quite regularly. Scientific chauvinism may have many sounds. When it is loud and overt, most of us condemn it quickly. But its subdued versions often go unnoticed and do a lot of harm.

Let me guide you through the different sounds of chauvinism in science, starting with loud examples and finishing with subtle ones. This didactic *decrescendo* will show you the full dynamic range of the problem and expose its often disregarded *pianissimo* versions.

Here is a *fortissimo* example that needs little discussion. A few years ago, the delegates from a Southeast European country refused to attend a biochemistry meeting with the argument that the meeting should have been held in *their* country. Such boycotts were common fare during the old Soviet days, but everyone knew that it was the politicians who foisted them on the scientists. This time, though, it was the scientists themselves who waved the banner of chauvinism.

The next example, a robust *forte*, is just as crude, but probably more common. The scene was the wood-paneled boardroom of a wealthy European foundation that had asked me, and several colleagues from around the world, for advice on how to dish out the foundation's fortunes. (If that sounds easy to you, just try it and you will quickly change your mind.) It had been a long day and we were finally free to choose our favorite drink and our favorite colleague (in that order), and were settling in to make small talk and let down our hair (to the extent it was still available.) The committee chairman had selected me as his favorite colleague and confidentially inquired 'whether those bloody Italians can ever do anything right'. When I suggested that they had produced a few acceptable violins and perhaps also one or two adequate paintings,

I quickly lost my favorite colleague status and had to look for someone else to talk to. This charming Italophile still runs a major scientific funding agency in his country and who knows what else.

Now we are down to a *mezzo piano*. The scene: a seminar room at a Swiss university. At the lectern: a seminar speaker. Seminar speaker (finishing his talk): 'Similar work has been reported by AB at Yale, by CD in Heidelberg, and also by some Japanese'. Audience: listens politely, nobody blinks. Presumably all Japanese are nameless little fellows who work and live in rabbit hutches. We have heard it before.

As our scale diminishes to a *piano*, it assumes a decidedly English accent. Creativity does not lend itself to statistics, but I would guess that at least two-thirds of the important discoveries in the life sciences now come from the United States and Great Britain. Why this is so need not concern us here. What matters is that the life sciences have become the playground of the English-speaking people and that their language has become the tyrant of today's molecular biology. Some West European countries are honorary club members as long as they play by the rules and do not speak a Latin language. Israelis are also in. Arabs are out. Indians, South Koreans and Chinese, whose scientific savvy is on the rise, are pounding on the door and are being considered for membership. If you want to make it in the life sciences, you had better speak decent English, dress the Western (preferably American) way, act 'cool', and select your literary quotes or jokes from the Anglophone repertoire. Pre-eminence is a fast road to self-importance and chauvinism. The fact that this chauvinism is often subconscious does not lessen the pain it can inflict on its victims. Here are two examples of this Anglophone chauvinism.

A few years ago, the preliminary program of a major international congress provoked an uproar because more than 90 percent of the slated speakers were from the USA or Great Britain. The program had been drafted by US scientists with impeccable professional and ethical credentials, but had to be hurriedly changed to forestall an international boycott.

In the same year, a prominent British science journal that likes to see itself as international featured a letter from an Asian scientist who wanted to know why the journal solicited the vast majority of its commentary articles from British or US scientists. I thought that the inquiry was amply justified, but the Editor's reply was a model of huffed surprise. Somebody should have told him that victims of chauvinism tend to have better antennas than chauvinists.

I could go on, but these two examples should do because they are so paradigmatic. Our US and British colleagues are rarely aware of their chauvinistic blinkers and usually discard them readily when made aware of them.

Artists, too, have sensitive antennas for scientific chauvinism. In his moving masterpiece *The Little Prince,* Antoine de Saint-Exupéry features a Turkish astronomer who tells an international audience about his discovery of a new asteroid. The audience ignores him because of his traditional Turkish garb, but when he dresses the Western way and presents the same findings again, they are accepted with enthusiasm. 'Grownups are really strange,' concludes the Little Prince.

When it comes to concocting nationally biased scientific programs, we scientists in Continental Europe are no slouches either. A few years ago, a European science agency reprimanded me because I had drawn up a meeting program in which too many speakers were from the United States and Great Britain. When I asked, perhaps a little pointedly 'What is the maximum tolerable Anglo-Saxon quota?' they backed off.

Chauvinism is a French word, so I must not forget to pay my respects to the French. A few decades ago, my French colleagues were stunned when their government ordered them to give all their lectures at international meetings in French. The orders were clear: No French, no travel money. *Quelle bêtise!* (The English translation starts with 'bull'.)

Citation bias, the tendency to ignore work done in other countries, is a particularly subtle form of scientific chauvinism because it is tricky to prove. This problem has been debated for decades, yet its existence is still not generally accepted. No wonder, because we have now reached the *pianissimo* end of our scale. My own experience has convinced me that a scientific discovery is more likely to be cited by others if it was made at a prestigious institution or in a scientifically prominent country. This form of scientific chauvinism inflicts great injustice and damage to researchers that work at less visible institutions or in disadvantaged countries. Citation chauvinism is so pernicious because its preferred victims are scientists who already have the odds stacked against them.

Doing science does not immunize us against chauvinism because doing science is not enough to make us scientists. If we want to be true scientists, then science must be more than just a profession to us. It should be a way to see ourselves and the world around us. It should be a yardstick for our daily actions and a beacon that tells us where to go.

Here is what this beacon tells me about scientific chauvinism: scientists are just human beings with their usual failings, so there will always be chauvinistic scientists. They are not the main problem. The main problem is the refusal to acknowledge that scientific chauvinism exists.

My ideal academic community is a sanctuary without intellectual taboos, where everything is open to reasoned and dispassionate discussion. Reality is different. There are topics an untenured assistant professor better avoid at faculty parties – and 'chauvinism in science' is one of them. I have stopped counting how often I got into hot water upon broaching this subject with colleagues. Many of them thought me overly sensitive, paranoid – or chauvinistic.

Such a denial posits that something does not exist because it is not supposed to exist. It is intellectual hypocrisy at its worst, an insult to the scientific spirit. Science insists that we see things as they are, unclouded by superstition, prejudice, official dogma – or political correctness. It forbids us to see the emperor's clothes if there are none, and commands us to say so clearly. Science, and particularly biology, has taught us that much of our social behavior is governed by primeval reflexes that are etched into our genes and expressed through our biochemical circuits.

I am convinced that some of these genetically determined circuits prompt us to reject people who are different. Such a trait may have served us well in our past as pack hunters, and has simply stuck around, like our appendix. How could we ever hope that choosing a particular profession would deliver us from this genetic appendage?

I wish we knew more about how our genetic programs shape our reaction towards others. This area is still uncharted territory that holds the promise of spectacular discoveries. In the late sixties, tantalizing glimpses of what may be ahead have come from the discovery that a few simple chemicals control the social behavior of certain insects. When these insects talk to each other, they use smells rather than sounds. And if they do not like each other's smell, they may try to kill each other.

Our brain is infinitely more complex than that of insects, yet still responds to some external chemical commands. Women living together in dormitories unwittingly synchronize their menstrual cycles by releasing volatile substances that are not consciously perceived as smells. And there may be other such volatile chemicals that control human aggression. Although these findings are still poorly researched and highly debated, they strongly suggest that chauvinism and racism are not social aberrations that can be stamped out, but inextricable threads in the fabric of our humanity.

Nothing suggests as yet that volatile chemical signals contribute to our chauvinistic or racist tendencies. Still, what little we know points to the fact that genetic differences influence social interactions between human beings. It would be foolish to negate this fact. We try to cover our ancient genetic circuits by layers of 'culture', and to some extent cultural silencing works. In fact, it is our only effective and acceptable defense. Yet it also inculcates our children with preconceptions and prejudices that can make these circuits even more dangerous. History tells us that these circuits may short out without warning and that each human being is a disaster waiting to happen.

A radiologist once told me that dark spots in the X-ray photos of my lungs attested to abortive attempts by *Mycobacterium tuberculosis* to invade my body when I was a child. My defenses had overwhelmed the intruders and imprisoned them in hardened tombs. But the unwelcome guests are still there, waiting for their chance. It would be foolish to ignore their presence. So I try to live a healthy life and keep my fingers crossed.

Maybe this is also good advice for handling the time bombs the world of my childhood has implanted into my brain. I am sure my spring-cleaning has not removed them all. I do not know where they are, or what they might do if they should go off. Looking at the world through a scientist's eye has helped me to put them away in protective shells, but I know that these shells are not perfect. At some limiting pressure, they will burst. What is this limiting pressure? Will I be lucky enough never to know? There is not much I can do about those bombs – except to keep reminding myself that they are there.

As long as science is done by humans, it will struggle with chauvinism. And the events of the past years suggest that the problem will get worse. Geopolitical tensions are on the rise and biological discoveries have become money machines, instruments of technological dominance, or substitutes for national flags. Science has taught us that we can only solve a problem by looking it squarely in the eye. Let's do so with scientific chauvinism – and then tell it to go away.

LETTING GO

A FEW YEARS AGO I RETIRED. I SAW TO IT that all people in my lab got jobs elsewhere, cleaned out my office, and walked into a new world. I was a little scared, because I had done research all my adult life, had done it with great passion, and did not know whether I was strong enough to go 'cold turkey'. My friends thought I had gone bonkers, because I could have stayed on for seven more years. Why give up research, the greatest game in town? Why not stay on as long as possible? After all, there were time-honored ways for pushing back that long vacation. Early retirement was shameful — a cop-out.

My decision to retire early had not come overnight, but had been building up over the years.

What could possibly beat being a university professor? I did not know — that was the point. I had to find out. What about giving more of me to my family and friends, or to music, books, writing, the 'small' things in life? The list was endless. There was so much left to do and the hour was getting late.

Also, I was no longer willing to put up with the constant rush of my profession. Between those deadlines to meet, those papers to publish, or those planes to catch, there was no longer time to think about what I was doing, why I was doing it, and what it all meant. Our crowded scientific meetings have always made me feel ill at ease, even when I was slated to give the opening lecture or receive an award. As a young assistant professor, I had often told my wife about my work, but there never seemed time for this anymore and she had stopped asking me about it long ago. Science should be a quiet conversation with nature, but I could no longer hear what she tried to tell me.

We are rarely sure why we do things, particularly if our motives are beyond logic and not easily put into words. Perhaps it was not only the noise. Perhaps I felt that there was something in me that my life in science had suppressed and that now wanted out. Was it my emotional self? I had always tried to keep it under wraps because our scientific profession considers emotions suspect, if not embarrassing. Perhaps the bright glare of science had made me miss the shade. Light helps us see things, but most of us think better in the shade.

If this is bunk to you, I am not offended. Giving up a job is a very personal matter, particularly if that job is as exciting and creative as that of an academic researcher. There is no general solution. Yet I strongly feel that, as a matter of principle, every professor should have to retire at a fixed age and make room for the next generation. But the university leadership should be free to make exceptions. Professors who have gone stale should be coaxed into early retirement — out of mercy for science and the students. And those rare individuals who keep going from strength to strength despite their age should be allowed, or even persuaded, to continue on a rolling contract for as long as they are productive. Limiting such contracts to 5 or 10 percent of all retiring professors would ensure that these exceptions remain just that, and also give the university leadership the all-important flexibility to do its job well.

But let's return to earth. As things stand now, we professors use every trick in the book to stay on longer. We 'cannot abandon our PhD students', even though we accepted them shortly before we were supposed to retire: the 'students-as-hostage' trick. We pull in large research grants and then use the overhead to soften our dean's resolve: the 'bribing-by-overhead' trick. We come up with grandiose plans for a new lecture course that only we could teach: the 'teaching-as-leverage' trick. Some of us torpedo the recruitment of a successor so we can squeeze out a few extra years by replacing ourselves: the 'I-am-indispensable' trick. Our colleagues in the USA have persuaded the courts to declare mandatory retirement illegal, and those in France have marched in the streets to protest against a government plan to make them retire earlier: the 'discrimination-based-on-age' trick. The list goes on, but you get the picture.

The discrimination-based-on-age trick disturbs me the most. It is selfish and scoffs at the unspoken covenant between generations. It is liberal democracy spun out of control. There is also the danger that countries other than the USA will fall for it; refusal to retire flies in the face of what we know about scientific creativity. Most scientists have their best ideas and make their most original discoveries while still young. And I am not even talking of mathematicians, whose golden age usually ends before they turn thirty. Young talent is our universities' life-blood and if this blood no longer flows freely, our universities are risking intellectual anemia. In the long term, refusal to retire will make academic tenure untenable. Tenure was meant to protect professors from arbitrary dismissal, not from mandatory retirement. Many outside the academic community already look at tenure with a jaundiced eye, because it has become a highly unusual privilege. At a time when top managers in the private sector must step down at an ever younger age, professors who refuse to retire at all are bizarre. They tie up precious faculty positions and expensive infrastructure and make our ivory tower even more ivory. They endanger us all. There are good reasons why the airlines retire their pilots, and there are equally good ones why universities should do the same with their professors.

Retirement need not be the end of a research career. Retired professors can join another laboratory as a guest, living off their retirement income, becoming long-term sabbatical visitors, as it were. Sabbatical visitors are nearly always a blessing to the host lab. I have hosted some twenty of them and one day I will try to sing their praise. They made my life less lonely, because I could use them as my Wailing Wall when I had difficulties with my research group. The members of my lab, in turn, valued these visitors as benevolent aunts and uncles who could defuse touchy situations and advise them when it was time to look for another job, or write a manuscript. All sides profited. In today's management parlance, it was a win-win-win situation.

I have often wondered why so few of us opt for this route. Perhaps it has to do with prestige and power. Both are acutely habit-forming and many older professors have become power junkies. Join another lab without calling the shots? No way! Much better to stop doing research and cling to the old office, the last vestige of past glory, and 'finally write up those experiments we did years ago' – immensely important experiments, no doubt. I also remember a grant application from someone who 'had just retired and now was finally ready to do research'.

Still others devote their newly gained time, their experience, and their personal contacts to giving long-winded speeches at faculty meetings, foisting unsolicited advice upon colleagues, deans and university presidents, or engaging in intrigues and back-door politics. Homer's Nestor stuck to his speeches, but these *Ersatz Nestors* also want to go into battle, armed with the phone and the department letterhead. Older professors are, of course, not the only ones playing academic power games, but they are more likely to use them for preserving the *status quo* – be it in university politics or in science. But science needs the free competition of ideas and when power distorts this competition, things quickly go awry. The history of science tells many sad stories of powerful older scientists who blocked scientific progress. Think of what Rudolf Virchow did to Robert Koch's ideas on the bacterial cause of tuberculosis, or how Richard Willstätter delayed general acceptance of James Sumner's discovery that enzymes are proteins.

Retired professors have still other options. Their life-long experience in teaching, research, and the inner workings of higher education makes them a valuable natural resource. Most of them have a worldwide network of acquaintances, colleagues and friends. And they have time. Who would be better qualified to run scientific organizations, professional journals, think tanks, university boards, or governmental advisory bodies?

Everywhere I look, science is under siege. Politicians want it to focus on trendy subjects and turn a quick profit. The public wants it to avoid any risk. And administrators manipulate it in a top-down fashion even though they have little idea of what science is, what it needs, and what it can or cannot do. If you don't believe me, visit Brussels and get a first-hand look at the schemes of our EU science administrators there. Or ask our US colleagues about what they have to deal with these days. In Europe, things got the way they are because Europe's best scientists either did not want to get involved in science politics, or because they were actively excluded from it. Perhaps top scientists should not do science politics at the peak of their research career, but what about those who have just retired? Their experience and prestige could help them deal with political and administrative decision makers and improve conditions for the next scientific generation. And retired professors can speak their mind freely, because they have nothing to fear. Tenured professors are well protected, but retired professors are unassailable – unless they want to publish old experiments.

But there is a big if: if one goes into science politics, the hands should still be warm from research. They usually cool quickly and then there is danger ahead. Someone who has been out of research for many years usually has lost touch with it and tends to make bad decisions. Here, too, it is important to know when to step down.

How to end is as important as how to begin, and at least as difficult. We biochemists have learned that the termination of a complex pathway is usually as intricate and as precisely controlled as the initial steps.

Our cells have designed an intricate machinery to start a new protein chain, but have gone to almost the same length to ensure that a chain ends at the right place. This principle even holds for the life of an entire cell. When it is time for a cell to go, it quietly cuts itself to pieces, wraps these into little membrane bags, and disappears without a fuss. The processes governing this programmed cell death appear to be as complex as those that govern growth. And when apoptosis fails, the results are malformed limbs, dementia, or cancer. Our own cells show us how to exit gracefully – why not learn from them?

In the final scene of Léos Janaĉek's enchanted opera *The Cunning Little Vixen*, an aging hunter returns to his forest after a long absence and asks the animals about their parents whom he used to know. When the young creatures tell him that they are already the grandchildren, he falls on his knees, overwhelmed by life's immutable flow. I am sometimes reminded of this finale when I pay a rare visit to my old institute and come across pieces of equipment or bottles whose label bears my handwriting. The young people who use them do not know me, nor do I know them, and most of them take me for a seminar visitor. When I chat with them about their background and their experiments, their optimism and enthusiasm always touch me. Retired professors, unlike hunters, do not easily fall on their knees, yet these visits always encourage me to continue on my present course. It has led me into uncharted waters where the noise has subsided and I am learning unexpected things. Now I know that life gives its riches to those who fight, but reserves its sweetness for those who can also let go.

EPILOGUE

I GOT HOOKED ON SCIENCE AS A BOY AND HAVE spent most of my life doing research. There were struggles, disappointments and failures, but overall it was a great party. Luck had something to do with it, because the second half of the twentieth century was a Golden Age for science. After the Great War, every year seemed better than the previous one and we saw no reason why this should ever stop. Three revolutions – the biological, the digital, and the sexual – opened up new worlds and made us see everything with new eyes. The Age of Reason seemed just around the corner.

Today the heady optimism and belief in progress that marked the Europe of my youth have given way to fundamentalism, intolerance and superstition. We have never needed science more than now. We need its discoveries, of course. We also need its dispassionate way of looking at the world and ourselves. Above all, however, we need its disrespect for dogmas, backward traditions, and the fanatical high ground.

But science is not invulnerable – it is like a tree. When it grows freely, nothing can resist it, but it withers quickly when we manipulate it and tell it what to do. The global discovery machine has become so big, so expensive, and so political that the tree of science shows signs of distress. We must take good care of this magic tree so that it will bear fruit for future generations.

Science prefers to give its choicest fruits to those who look for them in a playful way. It never gives us eternal truths, so we can only build models, knock them down, and then build better ones. This game goes on and on, with no end in sight. It may not be the ideal way to find out what the world is like; it may not even be the only way; but it is still one of the best I know. It has worked pretty well so far and, if done right, is a lot of fun. This fun is essential for science's health. When scientists withhold information and defend their models with clenched teeth, science suffers and it is time to worry. And I do worry now.

My years in science politics have made me aware of the winds that now blow into science's face. Writing these essays has propped up my spirit and reminded me what to fight for. No wonder some of these essays resemble a call to arms. Others pay tribute to the wacky humor that is so wonderful in our profession. And then there are some essays that sing the praise of life's chemical inventiveness.

At first sight, writing an essay seems easy. It is like writing a scientific paper without experimental results and there is nothing to curb your imagination – or your prejudices. But it shows your face as it is, because no editor will smoothen the wrinkles or remove the warts. And not having to prove anything makes it easy to lose track of where you are going. By the time you are halfway through, you realize that you are walking on a tightrope without a net, and you are scared.

Being scared is part of a scientist's life, because most experiments are expeditions into the unknown. Which of us has not sensed that butterfly in the stomach when starting a risky experiment? Michel Eyquem de Montaigne, too, must have known this sensation when he started his experiments (*Essais*) on himself. His genius allowed him to transform these experiments into one of the great books of humankind. Montaigne was not a scientist in the modern sense, but everything in his *Essais* conveys the liberating breeze of the scientific spirit. He showed us how to tend the tree of science and make it bear fruit. He was my unattainable North Star when, on my own *Vol de nuit,* I wrote the essays collected in this book.

ACKNOWLEDGMENTS

SOME OF THESE ESSAYS WERE BORN ON TRAINS, at my home, or on Greek shores, but most of them entered this world in a quiet library of the Institut Curie at Paris. I am very grateful to Dr. Daniel Louvard for having opened to me the doors of this little paradise, and to Madame Marie-Claude Moutier for being its friendly guardian angel. I am also indebted to Dr. Felix Wieland who proposed that I write this series of essays for FEBS Letters, suggested their present title, and helped me in many other ways by his keen intellect, humor and literary acuity. His editorial colleagues Patricia McCabe, Anne Müller and Tine Walma provided invaluable help in revising these essays for a more general audience and catching the many unintended results of my single-finger typing style.

Writing these essays showed me once again that many of my professional colleagues are also my friends who are willing to give me their time and their wisdom when I call upon them for help. These interactions sometimes refreshed a friendship that had struggled against many years and even more miles of separation. I am grateful to Markus Affolter, Christoph Dehio, Lisa and Fereydoun Djavadi, Stuart J. Edelstein, Susan Gasser, Benjamin S. Glick, Claus Kopp, Guido Kroemer, Chris Miller, Daniel Mojon, Uli Müller, Markus Noll, Dieter Oesterhelt, Ueli Schibler, John Spudich, Andreas Tammann, Steve Theg, Rüdiger Wehner, Wolfgang Wieser, William T. Wickner, Michael P. Yaffe and Charles Yocum for their comments and criticisms. Vidyanand Nanjundiah, editor of the Indian Journal of Biochemistry, kindly allowed me to publish a modified version of an essay which had originally appeared in that journal. Even my wife Merete and our daughter Isabella Seiler chipped in with valuable advice. Many of my friends have helped me with more than one article, but I hope they will forgive me if I single out those who have midwifed many, or even most of them: Heimo Brunetti, Sabeeha Merchant, and Michael P. Murphy.

Finally, I wish to thank Isabella Seiler and Greg Harris for the cover design, Kamilla Schatz and Ewald R. Weibel for the cover photos and "Pfuschi" (Heinz Pfister, Bern) for the cartoons.

Printed and bound by CPI Group (UK) Ltd, Croydon, CR0 4YY

13/10/2024

01773507-0001